Board Review Series

MICROBIOLOGY

AND

IMMUNOLOGY

Board Review Series

MICROBIOLOGY
AND
IMMUNOLOGY

ARTHUR G. JOHNSON, Ph.D.
RICHARD ZIEGLER, Ph.D.
THOMAS J. FITZGERALD, Ph.D.
OMELAN LUKASEWYCZ, Ph.D.
LOUISE HAWLEY, Ph.D.

Department of Medical Microbiology and Immunology
University of Minnesota
Duluth, Minnesota

SANS TACHE

WILLIAMS & WILKINS
Baltimore • Hong Kong • London • Sydney

Editor: Kim Kist
Associate Editor: Marjorie Kidd Keating
Copy Editor: Melissa Andrews
Design: Norman W. Och
Illustration Planning: Lorraine Wrzosek
Production: Barbara Felton

Copyright © 1989
Williams & Wilkins
428 East Preston Street
Baltimore, Maryland 21202, USA

Printed in the United States of America

Library of Congress Cataloging in Publication Data
Microbiology and Immunology / Arthur G. Johnson . . . [et al.].
    p. cm. — (Board review series)
    Includes index.
    ISBN 0-683-04466-4
    1. Medical microbiology—Outlines, syllabi, etc.   2. Medical microbiology—Examinations, questions, etc.   3. Immunology—Outlines, syllabi, etc.   4. Immunology—Examinations, questions, etc. I. Johnson, Arthur G. II. Series
    [DNLM: 1. Allergy and Immunology—examination questions.
2. Allergy and Immunology—outlines.   3. Microbiology—examination questions.   4. Microbiology—outlines.   QW 18 M6242]
QR46.M5387 1989
616'.01;076—dc19
DNLM/DLC
for Library of Congress   88-27681
                                                          CIP
                              89 90 91 92 93
                        1 2 3 4 5 6 7 8 9 10

PREFACE

There is an increasing quest among concerned educators for ways to transmit to medical and graduate students the enormous outpouring of relevant information from the basic and clinical laboratories of medical scientists. Dissemination of such information as an assemblage of current, important facts for recall following didactic courses is one such way. Hence this compendium has been formulated with the hope that it will serve as a useful and rapid recall of information relative to the microbial world. It includes basic characteristics and attributes of pathogenicity, as well as the multifaceted ways in which the human host has developed over eons to subvert the invasiveness of potential pathogens or to live synergistically with our normal flora.

This book is not meant to be a substitute for a microbiology textbook, of which there are many of excellence. Rather, it is a succinct listing of what we consider to be the most important concepts and facts applicable to the multiple, microscopic members of the microbial world and the ways in which the equilibrium between the host and parasite is affected. It covers the bacteria, viruses, and fungi as well as the immune system. Parasitology was deemed to be outside the scope of this effort, as was microbial genetics, which in the past two decades of research has resulted in such an enormous accumulation of information that it deserves a tome unto itself.

The book is composed of five major sections: General Properties of Microorganisms; Properties of Pathogenic Bacteria; Virology; The Fungi; and A Review of Immunology. Within each section the important general concepts shared as a group as well as the idiosyncrasies of the important organisms are accented. This format should promote rapid access to those organisms on which the student desires information. At appropriate junctions, pertinent questions are inserted, followed by the answers and explanations. These reflect the style characteristic of the National Board of Medical Examiners and should serve to highlight areas of importance.

The authors recognize that in publications in rapidly changing areas of science there will be omissions and shortcomings. The selection of information and the decisions made have been judgmental based on years of teaching medical and graduate students. We take full responsibility for any errors and appreciate their being drawn to our attention.

Much cooperative effort from diverse sources goes into a publication of this nature. The authors wish to thank Ms. Donna Kotter for effort above and beyond the call of duty. We also are appreciative of the invitation from John Gardner, Editor-in-Chief and Vice President, to give life to our dormant educational desires, as well as to Kim Kist and Margie Keating for their editorial assistance and for energizing us.

CONTENTS

5. A Review of Immunology 173

Index 231

1

General Properties of Microorganisms

The Microbial World

I. Microorganisms

–belong to the Protista biologic kingdom.
–include some eukaryotes, prokaryotes, viruses, viroids, and prions.
–are classified according to their structure, chemical composition, and biosynthetic and genetic organization.

II. Eukaryotic Cells

–contain organelles and a nucleus bounded by a nuclear membrane.
–contain complex phospholipids, sphingolipids, histones, and sterols.
–lack a cell wall (plant cells have a cellulose cell wall).
–have multiple diploid chromosome and nucleosomes.
–have relatively long-lived m-RNA with exons and introns.

III. Prokaryotic Cells

–have no organelles, nucleus, or histones, and only in rare cases complex phospholipids, sphingolipids, and sterols (mycoplasma).
–have a cell wall composed of peptidoglycan with muramic acid.
–are haploid with a single chromosome.
–have short-lived m-RNA with no exons.
–have coupled transcription and translation.

IV. Protozoa

–are eukaryotic cells.
–are classified into seven phyla; three phyla (Sarcomastigophora, Apicomplexa, and Cilophora) contain medically important species that are human parasites.
–are both commensal and pathogenic in human beings.

V. Fungi

–are eukaryotic cells.

–are divided into four major classes of which only one, the Deuteromycetes, causes human infections.

–can frequently exist as either a yeast (single cell) or a mold (multicellular structure); this phenotypic duality is called dimorphism.

–have both asexual and sexual reproduction.

–have a growth cycle that consists of both a vegetative and a reproductive phase.

VI. Bacteria

–are prokaryotic cells.

–are the smallest, free-living forms.

–may be normal flora or pathogenic in humans.

–do not have a sexual growth cycle, but some can produce asexual spores.

VII. Viruses

–are not cells and are not visible with the light microscope.

–are obligate intracellular parasites.

–contain no organelles or biosynthetic machinery except for a few enzymes.

–contain either ribonucleic acid (RNA) or deoxyribonucleic acid (DNA).

–are called *bacteriophage* or *phage* if they have a bacterial host.

VIII. Viroids

–are not cells and are not visible with the light microscope.

–are obligate intracellular parasites.

–are single-stranded, covalently closed circular RNA molecules that exist as base-paired rod-like structures.

–cause plant diseases, but have not been proved to cause human disease.

IX. Prions

–are infectious particles associated with scrapie, a degenerative central nervous system (CNS) disease of sheep.

–copurify with a specific protein and are resistant to nucleases but inactivated with proteases and other agents that inactivate proteins.

–are not fully understood in terms of their structure, replication, and pathogenesis.

Bacterial Structure

I. Bacterial Shape

–can usually be determined with appropriate staining and a light microscope.

–may be round (coccus), rod-like (bacillus), or spirals; cocci and bacilli often grow in doublets or chains.

–is used to identify bacteria.

–is determined by the mechanism of cell wall assembly.

–may be altered by antibiotics, which affect cell wall biosynthesis.

II. Bacterial Nucleus

–is not surrounded by a nuclear membrane nor does it contain a mitotic apparatus.

–may be observed in stained cells.

–consists of polyamine and magnesium ions bound to negatively charged, single-stranded DNA, small amounts of RNA, RNA polymerase, and other proteins.

III. Bacterial Cytoplasm

–contains ribosomes and various types of insoluble granules composed of compounds like poly-β-hydroxybutyric acid and volutin.
–contains no organelles.

IV. Bacterial Ribosomes

–have different proteins and RNAs than their eukaryotic counterparts.
–have a sedimentation coefficient of 70S and are composed of 30S and 50S subunits containing 16S, and 23S and 5S RNA, respectively.
–are the sites of action of many antibiotics that inhibit protein biosynthesis.
–are membrane-bound if engaged in protein biosynthesis.

V. Mesosomes

–are convoluted invaginations of the plasma membrane.
–function in DNA replication and cell division as well as secretion.
–are termed septal mesosomes if they occur at the septum (cross-wall) or lateral mesosomes if they are nonseptal.

VI. Cell Membrane

–is a typical unit membrane composed of phospholipids and proteins.
–contains the cytochromes and enzymes involved in electron transport and oxidative phosphorylation.
–is responsible for selective permeability and active transport, which are facilitated by membrane bound permeases, binding proteins, and various transport systems.
–contains carrier lipids and enzymes involved in cell wall biosynthesis.
–contains enzymes involved in phospholipid synthesis and DNA replication.
–contains chemoreceptors.
–is the site of action of certain antibiotics, e.g., polymyxin.

VII. Cell Envelope

–is composed of the macromolecular layers that surround the bacterium.
–always includes a cell membrane and a peptidoglycan layer.
–in gram-negative bacteria also includes an outer membrane layer.
–may include a capsule and/or a glycocalyx layer.
–contains components that frequently induce a specific antibody response.

A. Cell Wall

–refers to that portion of the cell envelope that is external to the cytoplasmic membrane and internal to the capsule or glycocalyx.
–confers osmotic protection and gram-staining characteristics.
–in gram-positive cells is composed of peptidoglycan, teichoic and teichuronic acids, and polysaccharides.
–in gram-negative bacteria is composed of peptidoglycan, lipoprotein, and outer phospholipid membrane, and lipopolysaccharide.

B. Peptidoglycan

–is also called *mucopeptide* and *murein*.
–is found in all bacterial cell walls.
–is a complex polymer consisting of (1) a backbone (alternating N-acetyl glucosamine and N-acetyl muramic acid) and (2) a set of identical tetrapeptide side chains, which vary from species to species, are attached to the N-acetyl muramic acid,

and are frequently linked to adjacent tetrapeptide by identical peptide cross-bridges that vary between species or by direct peptide bonds.

–comprises up to 50% of the cell wall of gram-positive bacteria, but only 2% to 10% of the cell wall of gram-negative bacteria.

–contains the β-1,4 glycosidic bond between N-acetyl muramic acid and N-acetyl glucosamine, which is cleaved by the bacteriolytic enzyme lysosyme.

–may contain *diaminopimelic acid,* an amino acid unique to prokaryotic cell walls.

–is the site of action of certain antibiotics, including penicillin, and cephalosporins.

C. Teichoic and Teichuronic Acids

–are found in gram-positive cell walls or membranes.

–are chemically bonded to peptidoglycan and/or membrane glycolipid, particularly in mesosomes.

–may account for 50% of the dry weight of a gram-positive cell wall.

–are water-soluble polymers containing a ribitol or glycerol residue linked by phosphodiester bonds.

–contain important bacterial surface antigenic determinants.

D. Lipoprotein

–cross-links the peptidoglycan and outer membrane in gram-negative bacteria.

–is linked to diaminopimelic acid residues of peptidoglycan tetrapeptide side chains by a peptide bond, and the lipid portion is noncovalently inserted into the outer membrane.

–anchors outer membrane to peptidoglycan.

E. Periplasmic Space

–is found in gram-negative cells.

–refers to the area between the cell membrane and the outer membrane.

–contains hydrated peptidoglycan, binding proteins, hydrolytic enzymes, and oligosaccharides.

F. Outer Membrane

–is found in gram-negative cells.

–is a phospholipid bilayer in which the phospholipids of the outer portion are replaced by lipopolysaccharides.

–protects cells from harmful enzymes and prevents leakage of periplasmic proteins.

–contains embedded proteins, including matrix *porins* (nonspecific pores), and some nonpore proteins (phospholipases and proteases), and transport proteins for small molecules.

G. Lipopolysaccharide (LPS)

–is found in gram-negative cells.

–consists of Lipid A (several long-chained fatty acids attached to phosphorylated glucosamine dissaccharide units) and a polysaccharide composed of a core and terminal repeating units.

–is attached to the outer membrane by noncovalent hydrophobic bonds.

–is negatively charged and noncovalently cross-bridged by divalent cations.

–is also called *endotoxin,* and the toxicity is associated with the Lipid A.

–contains major surface antigens including *"O" antigen* found in the polysaccharide component.

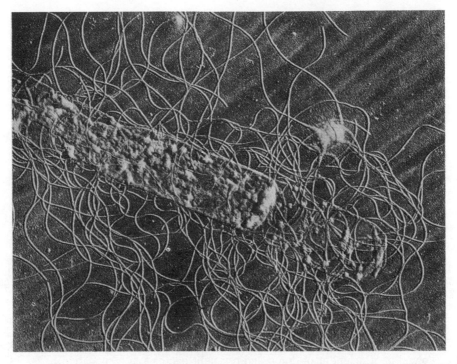

Figure 1.1. Shadowcast electronmicrograph of partly autolysed *Pr. vulgaris,* showing flagella, some of which appear to arise from spherical bodies remaining after autolysis (x 20,000). (Preparation by Dr. C. E. Robinow; from a micrograph kindly supplied by Dr. W. van Iterson.) (From Wilson GS, Miles A, eds. Topley and Wilson's Principles of Bacteriology, Virology and Immunity. 6th ed. Baltimore: Williams & Wilkins, 1975, vol. 1, p. 42.)

H. Capsule
–is a well-defined structure of polysaccharide surrounding a bacterial cell and external to the cell wall (Note: the one exception to polysaccharide is the poly-D-glutamic acid capsule of *Bacillus anthracis*).
–contributes to bacterial invasiveness because it protects bacteria from phagocytosis.

I. Glycocalyx
–refers to a loose network of polysaccharide fibrils that surround some bacterial cell walls.
–is sometimes called a *slime layer*.
–is associated with adhesive properties of the bacterial cell.
–is synthesized by surface enzymes.
–contains prominent antigenic sites.

VIII. Flagella
–are protein appendages for locomotion.
–are composed of a protein subunit called flagellin.
–may be located in only one area of a cell or over the entire bacterial cell surface.
–contain prominent antigenic determinants.

IX. Pili (Fimbriae)
–are rigid surface appendages composed mainly of protein.

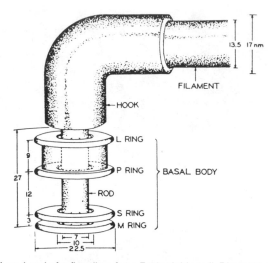

Figure 1.2. Model of the basal end of a flagellum from *Escherichia coli*. Dimensions are expressed in nanometers. (From DePamphilis ML, Adler J. Fine structure and isolation of the hook-basal body complex of flagella from *Escherichia coli* and *Bacillus subtilis*. *J Bacteriol* 1971; 105: 395.)

–exist in two classes: ordinary pili (involved in bacterial adherence) and sex pili (involved in attachment of donor and recipient bacteria in conjugation).

–are the *colonization antigens* associated with some bacterial species.

–contain the *M protein,* the major surface antigen of streptococci.

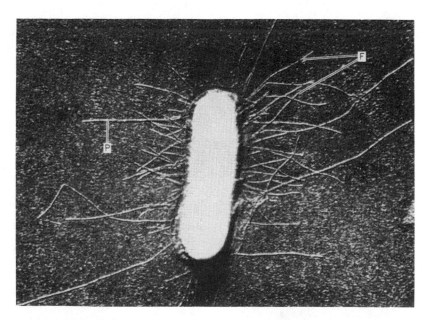

Figure 1.3. A shadowed preparation of *Escherichia coli* showing the fimbriae (F) and pili (P). (From Wilson GS, Dick HM, eds. Topley and Wilson's Principles of Bacteriology, Virology and Immunity. 7th ed. Baltimore: Williams & Wilkins, 1983, vol 1, p 32.)

X. Endospores

–are formed in response to certain adverse nutritional conditions.

–are inactive bacterial cells that are resistant to desiccation, heat, and various chemicals.

–possess a core that contains many cell components, a spore wall, a cortex, a coat, and an exosporium.

–contain calcium dipicolinate, which aids in heat resistance within the core.

–are formed in response to a bacterial differentiation process.

–germinate under favorable nutritional conditions following an activation process that involves damage to the spore coat.

–are helpful in identifying some species of bacteria.

Bacterial Growth and Cultivation

I. Bacterial Growth

–refers to an increase in bacterial cell numbers (multiplication), which results from a programmed increase in the biomass of the bacteria.

–requires a proper nutritional environment that allows the bacteria to reproduce.

–results from bacterial reproduction due to binary fission, which may be characterized by a parameter called *generation time* (the average time required for cell numbers to double).

–may be determined by measuring cell concentration (turbidity measurements or cell counting) or biomass density (dry weight or protein determinations).

–usually occurs asynchronously, i.e., all cells do not divide at precisely the same moment.

A. Cell Concentration

–may be measured by *viable cell counts* involving serial dilutions of sample followed by a determination of colony-forming units (CFU) on an agar surface.

–may be determined by *particle cell counting* or *turbidimetric density measurements* (includes both viable and nonviable cells).

B. Bacterial Growth Curve

–requires inoculation of bacteria from a saturated culture into fresh liquid media.

–is frequently illustrated in a plot of logarithmic number of bacteria versus time; the generation time is determined by observing the time necessary for the cells to double in number during the log phase of growth.

–consists of four phases: *lag* (metabolite depleted cells adapt to new environment), *exponential* or log (cell biomass synthesized at a constant rate), *stationary* (cells exhaust essential nutrients or accumulate toxic products), and *death* or *decline* (cells may die due to toxic products).

–is unique for a particular nutritional environment.

C. Growth Rate Constant (κ)

–is the rate at which bacterial cells are reproducing.

–may be determined by the formula $\kappa = \dfrac{Bdt}{dB}$ where B equals the biomass concentration and t equals time.

–is a function of the metabolic capabilities of the bacterial cell and its nutritional environment.

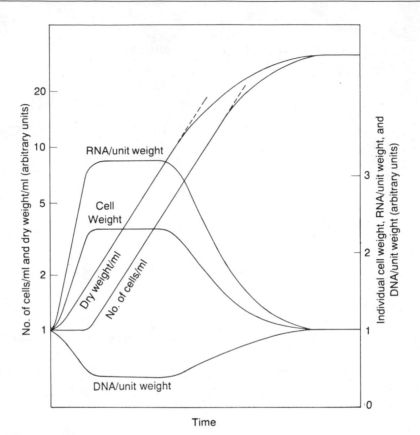

Figure 1.4. Diagram of changes in cell size and chemical composition during growth curve in length. Inoculum from an early stationary-phase culture. Bacterial count and dry-weight concentration on logarithmic scale (left-hand ordinate); average bacterial dry weight and content of ribonucleic and deoxyribonucleic acid on arithmetic scale (right-hand ordinate). Initial values of all variables taken as one unit (Herbert 1961a). (From Wilson GS, Miles A, eds. Topley and Wilson's Principles of Bacteriology, Virology and Immunity. 6th ed. Baltimore: Williams & Wilkins, 1975, vol 1, p 125.)

D. Chemostats or Turbidostats

  –are devices that maintain a bacterial culture in a specific phase of growth or at a specific cell concentration.
  –are most frequently employed to maintain a bacterial culture in the exponential growth phase.
  –are based on the principle that toxic products and cells are removed at the same rate as fresh nutrients are added and new cells synthesized.
  –operate best if one nutrient limits bacterial growth.

E. Synchronous Growth

  –refers to a situation in which all the bacteria in a culture divide at the same moment.
  –may be achieved by several methods, including thymidine starvation (thymidine-requiring bacteria), alternate cycles of low and optimal incubation temperatures, spore germination, selective filtration of older (large) and young (small) cells, or "trapped cell" filtration.

–can be achieved for only a few (one to five) cell division cycles.

II. Bacterial Cultivation

–refers to the propagation of bacteria.
–requires an environment that contains: (1) a carbon source, (2) a nitrogen source, (3) an energy source, (4) inorganic salts, (5) growth factors, and (6) hydrogen donors and acceptors.
–involves specific pH, gaseous and temperature preferences of bacteria.
–is done in either liquid (broth) or solid (agar) growth medium.

A. Obligate Aerobes

–refers to bacteria that require O_2 for growth.
–contain the enzyme superoxide dismutase, which protects them from the toxic free radical superoxide (O_2^-).

B. Obligate Anaerobes

–are killed by the superoxide (O_2^-) free radical.
–lack superoxide dismutase.
–require a substance other than oxygen (O_2) as a hydrogen acceptor during the generation of metabolic energy.
–use fermentation pathways with distinctive metabolic products.

C. Facultative Anaerobes

–will grow in the presence or absence of oxygen.
–shift from a fermentative to a respiratory metabolism in the presence of air.
–display the *Pasteur effect*, in which the energy needs of the cell are met by consuming less glucose under respiratory metabolism than under a fermentative metabolism.
–include most pathogenic bacteria.

D. Aerotolerant Anaerobes

–resemble facultative bacteria, but have a fermentative metabolism both with and without an oxygen environment.
–have the superoxide dismutase enzyme.

E. Superoxide Dismutase

–is an enzyme in aerobes and aerotolerant anaerobes that allows them to grow in the presence of the superoxide (O_2^-) free radical.
–carries out the following reaction: $2O_2^- + 2H^+ \rightarrow H_2O_2 + O_2$.
–produces H_2O_2 (peroxide), which is toxic to cells but is destroyed by catalase or oxidized by a *peroxidase enzyme*.

F. Mesophile

–refers to bacteria that grow in the temperature range of 20 to 45°C.
–may be distinguished from a *psychophile* (temperature range of 0 to 20°C) and a *thermophile* (temperature range of 45 to 90°C).
–is the temperature grouping of medically important bacteria.

G. Heterotroph

–refers to bacteria that require preformed organic compounds, e.g., sugar, amino acids for growth.
–may be distinguished from an *autotroph*, which does not require organic compounds because it can synthesize them from inorganic compounds.

H. Minimal Essential Growth Medium

–is a bacterial growth medium that contains only the primary precursor compounds essential for growth.
–demands that a bacterium synthesize most of the organic compounds required for its growth.
–dictates a relatively slow generation time.

I. Complex Growth Medium

–is a bacterial growth medium that contains most of the organic compound building blocks, e.g., sugars, amino acids, nucleotides, necessary for growth.
–may contain components like peptones, cell hydrolysates.
–dictates a faster generation time for a bacterium relative to its generation time in minimal essential medium.
–is necessary for the growth of fastidious bacteria.

J. Differential Growth Medium

–is a bacterial growth medium containing a combination of nutrients and pH indicators to differentiate visually bacteria that grow on or in it.
–is frequently a solid media on which colonies of particular bacterial species have a distinctive color.
–is a type of enrichment culture medium.

K. Selective Growth Medium

–is a bacterial growth medium that contains compounds that prevent the growth of some bacteria while allowing the growth of other bacteria.
–use certain dyes or sugars, high salt concentration, or pH to achieve their selectivity.
–is a type of enrichment culture medium.

Bacterial Metabolism

I. General Features

A. Bacterial Metabolism

–refers to all the chemical processes in a bacterial cell.
–is the sum of *anabolic processes* (synthesis of cellular constituents requiring energy) and *catabolic processes* (breakdown of cellular constituents with concomitant release of waste products and energy-rich compounds).
–is *heterotrophic* for pathogenic bacteria.
–varies depending on the nutritional environment.
–is usually dependent on facilitated diffusion, active transport, and group translocations of nutrients and passive diffusion of gases and ions into cells.

B. Bacterial Transport Systems

–involve membrane associated binding or transport proteins.
–frequently require energy to concentrate substrates inside the cell.
–are used to transport sugars and amino acids.
–are usually inducible for nutrients that are catabolized (glucose, which is constitutive, is an exception).
–use phosphotransferase systems frequently when sugars are transported.

C. Bacterial Energy Metabolism

–can occur in one of three types: fermentation, respiration, or photosynthesis.

–involves the oxidation of organic compounds (fermentation or respiration) in all pathogens.

II. Type of Metabolism

A. Fermentation

–is one of the methods by which some bacteria obtain metabolic energy.

–is characterized by a substrate phosphorylation.

–involves the formation of adenosine triphosphate (ATP) not coupled to electron transfer.

–requires an intermediate product of glucose metabolism (often pyruvate) as final hydrogen accepter.

–results in the synthesis of specific metabolic end products that aid in the identification of bacterial species.

–frequently involves the Embden-Meyerhof pathway.

–may result in formation of one end product (*homofermentation*) or several end products (*heterofermentation*).

B. Respiration

–refers to the method of obtaining metabolic energy that involves an oxidative phosphorylation.

–involves the formation of ATP during electron transfer and the reduction of gaseous O_2.

–involves a cell membrane electron transport chain composed of cytochrome enzymes, lipid cofactors, and coupling factors.

C. Photosynthesis

–involves a process known as cyclic photophosphorylation.

–is similar to respiration except that photochemical processes using energy of light are responsible for the synthesis of the reductant.

–does not occur in any medically important bacteria.

III. Carbohydrate Metabolism

A. Aerobic Carbohydrate Metabolism

–involves a combination of the Embden-Meyerhof pathway and the tricarboxylic acid cycle for most pathogenic bacteria; occasionally the Entner-Doudoroff pathway hexose monophospholic shunt is used instead of the Embden-Meyerhof pathway.

–is regulated by the levels of enzyme activity and enzyme synthesis.

–involves inducible, repressible, and constitutive enzymes.

B. Regulation of Enzyme Activity

–may occur because enzymes are *allosteric proteins* susceptible to binding of effector molecules that influence their activity.

–may occur by *feedback inhibition* due to end product.

–may involve substrate binding enhancement (*cooperatively*) of catalytic activity.

C. Regulation of Enzyme Synthesis

–may involve allosteric regulatory proteins that activate (activators) or inhibit (repressors) gene transcription.

–may involve *end product feedback repression* of biosynthetic pathway enzymes.
–may involve *substrate induction* of catabolic enzymes.
–may involve *attenuation control sequences* in enzyme m-RNA.
–may involve the process of *catabolite repression,* which is under positive control of the *catabolite activator protein (CAP).*

Sterilization and Disinfection

I. Definitions

–sterility: total absence of viable microorganisms as assessed by no growth on any medium.
–bactericidal: killing of bacteria.
–bacteriostatic: inhibition of growth of bacteria.
–sterilization: removal or killing of all microorganisms.
–disinfection: removal or killing of disease-causing microorganisms.
–sepsis: infection.
–aseptic: without infection.
–antisepsis: topical application inhibiting the growth and multiplication of microorganisms.

II. Kinetics of Killing

–affected by menstruum or media, concentration of organisms and antimicrobial agent, temperature, pH, presence of spores.
–can be exponential or logarithmic.
–killing curve can become asymptotic requiring extra considerations in killing final numbers, especially if population is heterogeneous relative to sensitivity.

III. Antimicrobial Agents

–moist heat; autoclaving at 121°C for 15 minutes at a steam pressure of 15 pounds per square inch kills microorganisms, including spore formers.
–ultraviolet radiation; blocks replication of DNA.
–chemicals.

A. Phenol

–used as a disinfectant standard.
–expressed as a phenol coefficient: the rates of the minimal sterilizing concentration of phenol to that of the compound.

B. Iodine

–bactericidal in a 2% solution of aqueous alcohol containing potassium iodide.
–acts as an oxidizing agent combining irreversibly with proteins.

C. Chlorine

–inactivates microorganisms by oxidizing free sulfhydryl groups.

D. Formaldehyde

–used as a disinfectant in aqueous solution (37%).

E. Ethylene Oxide

–an alkylating agent especially useful for disinfecting hospital instruments.

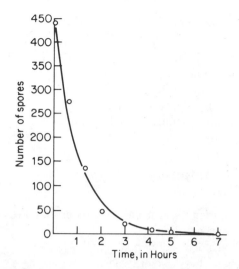

Figure 1.5. Disinfection of anthrax spores with 5 percent phenol at 33.3°. The curve is drawn through a series of calculated points: the circles represent the experimental observations. (From Wilson GS, Dick HM, eds. Topley and Wilson's Principles of Bacteriology, Virology and Immunity. 7th ed. Baltimore: Williams & Wilkins, 1983, vol 1, p 85.)

Antimicrobial Chemotherapy

I. Antimicrobial Chemotherapy

–is based on the principle of *selective toxicity*, which implies a compound is harmful to a microorganism but innocuous to its host.

–involves drugs that (1) are antimetabolites, (2) inhibit cell wall biosynthesis, (3) inhibit protein synthesis, (4) inhibit nucleic acid synthesis, and (5) alter or inhibit cell membrane permeability or transport.

–includes both *bacteristatic* (inhibit growth) and *bactericidal* (kill) drugs.

–may employ *synergistic* combinations of drugs.

–must consider both *drug-parasite relationships,* e.g., location of bacteria and drug distribution, and alterations of *host-parasite relationships*, e.g., immune response and microbial flora.

A. Drug Antimicrobial Activity

–is usually determined by *dilution* or *diffusion* tests.

–may differ in vitro and in vivo.

–is affected by pH, drug stability, microbial environment, number of microorganisms present, length of incubation with drug, and metabolic activity of microorganisms.

–may be modified for a specific bacterium by its development of *genetic or nongenetic drug resistance.*

B. Drug Resistance

1. Nongenetic Mechanism of Drug Resistance

–may involve loss of specific target structures, e.g., cell wall by L-forms of bacteria.

–may result from metabolic inactivity of microorganisms.

2. Genetic Mechanisms of Drug Resistance

–may result from either *chromosomal* or *extrachromosomal* resistance.

–may involve a chromosomal mutation that alters the structure of the drug's receptor or the drug's permeability.

–may result from the introduction of a plasmid that codes for enzymes that modify the drug by merely breaking bonds (β-lactamase) or by adding or subtracting functional groups (acetyltransferase).

–is frequently the result of the introduction of a resistance transfer factor (RTF) plasmid into a bacterium.

II. Mechanisms of Action

A. Antimetabolites

–include *bacteristatic* (sulfonamide, trimethoprim, and para-aminosalicylic acid) and *bactericidal* (isoniazid) drugs.

–are structural analogues of normal metabolites.

–inhibit the action of specific enzymes.

B. Cell Wall Synthesis Inhibitors

–are bactericidal.

–may inhibit transpeptidation (*pencillins and cephalosporins*; β-lactam drugs).

–may inhibit the synthesis of peptidoglycan (*cyclosporine, bacitracin, vancomycin, and ristocetin*).

–may act in the cytoplasm (*cycloserine*), or the membrane (*bacitracin*), or in the cell wall (*penicillins, cephalosporins, vancomycin*, and *ristocetin*).

–may cause bacteria to "take on" aberrant shapes or become *spheroplasts*.

1. Penicillins

–inhibit the transpeptidation enzymes involved in cell wall biosynthesis.

–are more active against gram-positive than gram-negative bacteria.

–react with penicillin-binding proteins (PBPs).

–are inactivated by β-lactamases (penicillinases) that are genetically coded in some bacterial DNA or some resistance-transfer factor plasmids.

2. Cephalosporins

–have a mechanism of action similar to penicillin.

–possess activity against both gram-positive and gram-negative bacteria.

–are inactivated by some β-lactamases.

–are frequently used to treat patients allergic to penicillins.

C. Cell Membrane Inhibitors

–alter the osmotic properties of the plasma membrane (*polymyxin and polyenes*) or inhibit membrane lipid synthesis (*miconazole and ketoconazole*).

–are *bactericidal*.

–are useful for some gram-negative (*polymyxin*) and sterol-containing mycoplasma and fungal (*nystatin and amphotericin B*) infections.

–can react with mammalian cell membranes and are therefore toxic; mainly used topically or with very severe infections.

D. Protein Synthesis Inhibitors

–include the *aminoglycosides* (streptomycin, neomycin), *tetracyclines, chloramphenicol, macrolides* (erythromycin), and *lincomycins* (lincomycin and clindamycin).

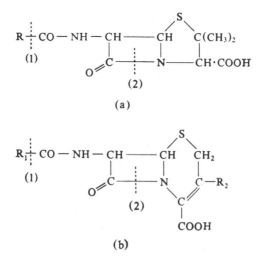

Figure 1.6. General formula of a penicillin (a) and a cephalosporin (b). (1) Site of action of acylases (splitting the amide linkage). (2) Site of action of β-lactamases. (From Wilson GS, Miles A, eds. Topley and Wilson's Principles of Bacteriology, Virology and Immunity. 6th ed. Baltimore: Williams & Wilkins, 1975, vol 1, p 212.)

–are *bactericidal* (aminoglycosides for gram-negative bacteria) or *bacteristatic* (tetracyclines, chloramphenicol, macrolides, and lincomycins).
–bind to either the 30S (aminoglycosides and tetracyclines) or the 50S (chloramphenicol, macrolides, and lincomycins) ribosomal subunits.
–are frequently known as *broad-spectrum* antibiotics.

1. Aminoglycosides

–include streptomycin, neomycin, kanamycin, and gentamycin.
–are *bactericidal*, bind to the 30S ribosomal subunit, and irreversibly block initiation of translation and/or cause m-RNA misreading.
–have a narrow effective concentration range before toxicity occurs, causing renal and eighth cranial nerve (hearing loss) damage.
–may be modified (acetylation) and rendered inactive by enzymes contained in resistance transfer factor plasmids.

2. Tetracyclines

–include tetracycline, oxytetracycline, chlortetracycline.
–are *bacteristatic*, bind to 30S ribosomal subunit, and prevent binding of aminoacyl t-RNA to accepter site.
–may be deposited in teeth and bones, which can cause staining and/or structural problems in children.
–are not transported into cells containing specific tetracycline resistance transfer factor plasmids.

3. Chloramphenicol

–is *bacteristatic* for gram-positive and gram-negative bacteria, rickettsia, and chlamydia.
–binds to the 50S ribosomal subunit and inhibits peptide bond formation.
–may be inactivated by the enzyme chloramphenicol acetyl-transferase, which is carried on a resistance transfer factor plasmid.

4. Macrolides and Lincomycins

–include erythromycin (macrolide) and lincomycin and clindamycin (lincomycins).
–are *bacteristatic*.
–bind to the 23S RNA in the 50S ribosomal subunit, and block translocation.
–are rendered ineffective in bacteria that have a mutation in a 50S ribosomal protein or contain a resistance transfer factor plasmid that possesses genetic information that results in methylation of 23S RNA, which inhibits drug binding.

E. Nucleic Acid Synthesis Inhibitors

–may inhibit DNA (*actinomycin, mitomycin, and nalidixic acid*) or RNA (*rifampin*) synthesis.
–are *bactericidal* and are considerably toxic to mammalian cells.
–bind to strands of DNA (actinomycin and mitomycin) or inhibit replication enzymes (rifampin—DNA dependent RNA polymerase, and nalidixic acid—DNA gyrase).

F. Other

1. Griseofulvin

–is a *fungistatic* drug active against fungi with chitin in their cell walls.
–inhibits protein assembly, which interferes with cell division by blocking microtubule assembly.

Toxins

I. Toxins are broadly defined as microbial products that damage host cells or host tissues. The older classical differentiation between exotoxins and endotoxins follows:

Property	Exotoxin	Endotoxin
organisms	gram-positive	gram-negative
composition	proteins	lipopolysaccharides
released by organisms	yes	no
heat sensitivity	labile	stable
toxoids (vaccines)	yes	no
neutralization by antitoxin	yes	no
degree of toxicity	very potent	less potent
specificity for target cells	high	low

Recently, a variety of enterotoxins have been identified. They exhibit characteristics of both exotoxins and endotoxins.

II. Many toxins possess an A and a B fragment. The B fragment is responsible for attachment of the toxin to the specific target tissue. The A fragment enters the cell and exerts its toxic effect. Antitoxin will interact only with the B fragment to block its attachment; once toxin is bound, the antitoxin is ineffective. At least five separate toxins containing these two fragments have been defined.

A. *Corynebacterium dipththeriae:* diphtheria exotoxin inhibits protein synthesis via the t-RNA EF-2 elongating factor.
B. *Pseudomonas aeruginosa:* exotoxin A also inhibits protein synthesis via the t-RNA EF-2 factor.
C. *Shigella dysenteriae:* shiga neurotoxin inhibits protein synthesis via the 60S ribosomal unit.

D. *Vibrio cholerae:* choleragen enterotoxin stimulates adenyl cyclase to overproduce cyclic adenosine monophosphate (AMP) and induce loss of fluids and electrolytes.
E. *Escherichia coli:* heat labile enterotoxin also stimulates adenyl cyclase to overproduce cyclic AMP and induce loss of fluids and electrolytes.

III. Other well-characterized toxins do not appear to have separate A and B fragments.

A. *Clostridium tetani:* tetanospasm exotoxin acts on synaptosomes; gangliosides bind the toxin and obliterate the inhibitory reflex response of nerves causing uncontrolled impulses (hyperreflexia of skeletal muscles).
B. *Clostridium botulinum:* botulinum exotoxin acts on myoneural junctions; cholinergic nerve fibers are paralyzed, which suppresses the release of acetycholine.
C. *Clostridium perfringens:* α-toxin is a phospholipase C; it disrupts cellular and mitochondrial membranes.
D. *Escherichia coli:* heat stable enterotoxin stimulates guanylate cyclase to overproduce cyclic guanosine monophosphate (GMP), which impairs chloride and sodium absorption.
E. *Salmonella:* enterotoxin stimulates cyclic AMP.
F. *Staphylococcus aureus:* exfoliative toxin disrupts the stratum guanulosum in the epidermis.

IV. In general, all pathogenic bacteria possess a variety of toxic capabilities. All of the above mentioned examples have been characterized in specific terms of disease induction. Numerous other toxins have been documented, but their precise role in pathogenesis remains to be determined.

Bacteriophages

I. Bacteriophages

–are bacterial viruses frequently called *phages.*
–are obligate intracellular parasites.
–are infectious agents for bacteria.
–are very "host-specific."

A. Bacteriophage Virions

–are the infectious phage particles.
–contain only protein and RNA or DNA as major components, although some lipid is present in some phages.
–have been morphologically divided into three classes: polyhedral, filamentous, and complex.

1. Polyhedral Phages

–are usually composed of an outer polyhedral-shaped protein coat (*capsid*), which surrounds the nucleic acid.
–may contain a lipid bilayer between two protein capsid layers (PM-2 phage).
–have either circular double-stranded (PM-2) or single-stranded DNA (ØX-174 and M-12) or linear single-stranded RNA (MS2 and Qβ) as their genetic material although one phage (Ø6) that has three pieces of double-stranded RNA has been described.

2. Filamentous Phages

–have a filamentous protein capsid that surrounds a circular single-stranded DNA genome (f1 and M13).
–are "male" bacteria-specific in that infection occurs through the pili, which are only present on "male" bacteria.
–do not lyse their host cells during the replication process.
–have a different replication pattern than other phage since their entire virlon penetrates the cell wall and enters the host [see II below].

3. Complex Phages

–have a protein polyhedral head containing linear double-stranded DNA, and a protein tail and other appendages.
–include the "T"- and λ-phages of *E. coli.*

4. RNA Phages

–refers to all phages with RNA as genetic material.
–are all specific for bacteria with male pili (*male specific*).
–contain single-stranded RNA (except for Ø6 noted above), which can act as polycistronic m-RNA.

5. DNA Phages

–refers to all phages with DNA as genetic material.
–may contain some unusual nucleic acid bases like 5-hydroxymethyl cytosine or 5-hydroxymethyl uracil.
–contain nucleic acid bases that are frequently glucosylated or methylated.
–are classified as *virulent* or *temperate* depending on whether their pattern of replication is strictly lytic (virulent) or alternates between lytic and lysogenic (temperate) [see II below].

II. Bacteriophage Replication

–requires that the phage use the biosynthetic machinery of the host cell.
–follows a basic sequence of events, which includes (1) *adsorption*, (2) *penetration*, (3) *phage-specific transcription and/or translation*, (4) *assembly*, and (5) *release.*
–is initiated by interaction of phage receptors and specific bacterial surface receptor sites.
–involves the injection of the phage genome into the host cell (filamentous phages are the exception).
–follows one of two types of patterns (lytic or lysogenic) for DNA phages.
–is usually complete in 30 to 60 minutes for virulent phages.

A. Lytic Replication Cycle

–occurs with virulent phage (*E. coli* "T"-phages).
–results in the lysis of the host cell.
–is a very inefficient process, but still leads to the production of 100 to 300 infectious phages per infecting phage.
–may be studied in an experimental situation that may be analyzed in a *one-step growth curve.*
–is the basis for *phage-typing* of bacteria.

1. **One-Step Growth Curve**

    –is the result of an experimental situation used to monitor one cycle of phage replication during lytic phage replication.

    –is a plot of infectious virus produced versus time after infection.

    –involves the use of an infectious center assay called a *plaque assay* (counts of focal areas of phage-induced lysis on a lawn of bacteria).

    –allows one to calculate the average time necessary for a phage to replicate within a specific host cell and be released from that cell (*replication time*).

    –allows one to determine the number of infectious phages produced from each infecting phage (*burst size*).

    –allows one to determine the *eclipse period* (time from infection to the synthesis of the first intracellular infectious virus).

2. **Phage Typing**

    –is a method used to identify strains of bacteria, e.g., *Staphylococcus aureus*.

    –is based on the lysis of the bacteria by a selected set of phages.

B. **Lysogenic Replication Cycle**

    –may occur *only* with temperate phages (*E. coli* λ-phage).

    –involves limited phage-specific protein synthesis due to the synthesis of a phage-specific *repressor protein* that inhibits phage-specific transcription.

    –includes the incorporation of *prophage* (the phage DNA) into specific attachment sites in the host cell DNA.

    –confers immunity to infection by a similar type phage.

    –results in the prophage being passed on to succeeding generations of the bacteria.

    –can revert to the lytic replication cycle if the phage repressor is destroyed.

    –can result in the generation of *specialized or restricted transducing phages*.

    –may result in the process known as *lysogenic phage conversion*.

1. **Lysogenic Phage Conversion**

    –refers to a change in the phenotype of bacteria due to the limited expression of genes within a prophage.

    –occurs in *Salmonella* polysaccharides due to the ε-prophage.

    –is the genetic mechanism by which nontoxigenic strains of *Corynebacterium diphtheriae* are converted to toxin-producing strains.

    –results in the conversion of nontoxigenic *Clostridium botulinum* types C and D to toxin-producing strains.

Review Test
BASIC MICROBIOLOGY

DIRECTIONS: Each of the questions or incomplete statements below is followed by suggested answers or completions. Select the *one best* in each case.

1.1. Which of the enzymes listed below would be most likely to affect sugar transport in bacteria?

A. Acetyltransferase.
B. Neuraminidase.
C. Oxidase.
D. Phosphotransferase.
E. Aminopeptidase.

1.2. If diaminopimelic acid is removed from the growth media of a bacteria that requires it, which structure or macromolecule would be affected?

A. Teichoic acid.
B. Cell wall.
C. Glycocalyx.
D. Lipopolysaccharide.
E. Capsule.

1.3. The bacterial structure most involved in adherence is the

A. Capsule.
B. Lipopolysaccharide.
C. Ordinary pili.
D. "O" specific side chain.
E. Flagella.

1.4. Cephalosporin antibiotics

A. are used to treat fungal infections.

B. are bacteriostatic.
C. inhibit protein biosynthesis.
D. are used to treat patients allergic to penicillin.
E. must reach the cytoplasm to be effective.

1.5. Bacteria that contain the superoxide dismutase enzyme

A. produce peroxide (H_2O_2 from H^+ ion and the superoxide radical O_2^-.
B. need superoxide to grow.
C. are frequently obligate anaerobes.
D. depend on O_2 as an energy source.
E. use their products as an energy source under fermentative conditions.

1.6. Lysogenic phage conversion refers to

A. the transformation of a virulent phage to a lysogenic phage.
B. a change in bacterial phenotype due to the presence of a prophage.
C. the conversion of a prophage to a temperate phage.
D. the incorporation of a prophage into the bacterial chromosome.
E. the immunity to infection that a lysogenic phage confers to a bacterium.

DIRECTIONS: Each set of lettered headings below is followed by a list of numbered words or phrases. Identify the lettered word most closely associated with the numbered phrase.

A. Viruses
B. Viroids
C. Both A and B
D. Neither A nor B

1.7. Are obligate intracellular parasites.
1.8. Contain RNA or DNA.

1.9. Cause human diseases.
1.10. Contain membrane-bound ribosomes active in protein synthesis.
1.11. Are dimorphic.

A. Cell wall synthesis
    inhibitor

20

B. Cell membrane inhibitors
C. Both
D. Neither

1.12. Include the β-lactam drugs.
1.13. Are bactericidal.
1.14. Are effective against bacteria in the stationary phase of growth
1.15. Are resistant to inactivation by the products of known resistance transfer factors.
1.16. Are effective against gram-positive bacteria.

A. Differential growth media
B. Selective growth media
C. Both
D. Neither

1.17. Allows the growth of all aerobic bacteria.
1.18. Is an enrichment culture media.
1.19. Allows one to distinguish between mesophiles and thermophiles.
1.20. Inhibits the growth of some bacteria.
1.21. Is based on the principle that different bacteria have different growth rate constants.

A. Gram-positive bacteria
B. Gram-negative bacteria
C. Both
D. Neither

1.22. Contain teichoic acids.
1.23. Have a cell wall that is composed of up to 50% glycocalyx.
1.24. Contain N-acetylmuramic acid.
1.25. Contain Lipid A.
1.26. Contain porins in their outer membrane.

A. Fermentation metabolism
B. Respiration
C. Both
D. Neither

1.27. Involves the oxidation of organic compounds.
1.28. Includes a process known as cyclic photophosphorylation.
1.29. Can occur entirely in the bacterial cytoplasm.
1.30. Occurs in the absence of O_2.
1.31. Involves the formation of ATP.

A. Virulent phage
B. Temperate phage
C. Both
D. Neither

1.32. Synthesize a repressor protein.
1.33. Form restricted transducing phage.
1.34. May attach to male pili to initiate infection.
1.35. Lyse infected cells when they are released.
1.36. Are used in phage typing.

DIRECTIONS: For each of the questions or incomplete statements below, *one* or *more* of the answers or completions given is correct. Choose answer:
A. if only **1, 2,** and **3** are correct
B. if only **1** and **3** are correct
C. if **2** and **4** are correct
D. if only **4** is correct
E. if all are correct

1.37. The bacterial growth curve for a certain strain of bacteria
1. is determined by the nutrients in the media.
2. is best measured during growth in a turbidostat.
3. depends on the enzymatic capabilities of the bacteria.
4. is likely to be similar in minimal and complex growth media.

1.38. Bacteristatic antibiotics
1. are a good choice to use in combination with aminoglycosides in order to obtain a synergistic effect.
2. include all the antimetabolites.

3. are effective against bacteria in all phases of growth.
4. include some "broad-spectrum" antibiotics.

1.39. The outer membrane
1. contains transport proteins.
2. is attached to "endotoxin."
3. is a structure found in gram-negative bacteria.
4. contains the enzymes involved in bacterial oxidative phosphorylation.

1.40. Prokaryotic cells
1. have coupled transcription and translation.
2. have spliced m-RNAs.
3. are generally haploid.

4. transfer secretory proteins into the Golgi apparatus before secretion.

1.41. The regulation of enzyme activity in bacterial cells

1. can be coupled to the binding of allosteric proteins.
2. can be controlled by a catabolite activator protein (CAP).
3. may occur via feedback inhibition.
4. can involve inducer molecules.

Identify the lettered phrase related most closely to the numbered statement.

A. *Escherichia coli* heat labile toxin
B. *Vibrio cholerae* choleragen
C. *Corynebacterium diphtheriae* exotoxin
D. *Pseudomonas aeruginosa* exotoxin A
E. all are correct
F. none are correct

1.42. A fragment is toxic; B fragment is responsible for attachment to host tissue or cell.

1.43. can also be classified as an endotoxin.

1.44. stimulates guanylate cyclase to overproduce cyclic GMP.

1.45. acts at myoneural junction to paralyze cholinergic nerve fibers.

1.46. phospholipase C that disrupts mitochondrial and cell membranes.

1.47. causes a type of food poisoning.

Answers and Explanations

BASIC MICROBIOLOGY

1.1. D. The transport of sugar into a bacterium frequently involves the transfer of a phosphate group to the sugar molecule.

1.2. B. Diaminopimelic acid is an amino acid that is a component of the mucopeptide portion of the cell wall of some bacteria.

1.3. C. Ordinary pili and glycocalyx are the two bacterial structures that are involved in adhesiveness.

1.4. D. Patients who are allergic to the various penicillins are frequently given one of the cephalosporins.

1.5. A. Superoxide dismutase is found in aerobic bacteria and protects them from the toxic free radical $(O_2{}^-)$ by combining it with H^+ ion to form H_2O_2, which is subsequently degraded by peroxidase.

1.6. B. This term refers to a change in bacterial phenotype due to the presence of a lysogenic prophage of a temperate phage.

1.7. C. Viruses and viroids must infect cells in order to multiply.

1.8. A. Viruses contain either RNA or DNA; viroids contain only single-stranded RNA.

1.9. A. Viroids have only been proved to cause plant disease.

1.10. D. Viruses and viroids must use the ribosomes of host cells for protein synthesis.

1.11. D. Fungi are dimorphic, i.e., have two morphologic forms.

1.12. A. The β-lactam drugs are those antibiotics like the various penicillins and cephalosporins that contain a β-lactam ring in their structure.

1.13. C. Antibiotics of both types kill bacteria.

1.14. B. Bacteria in the stationary phase of growth are not synthesizing cell wall, therefore only cell membrane inhibitors that don't require ongoing membrane biosynthesis are effective.

1.15. B. There are no known resistance transfer factor products that affect cell membrane inhibitors.

1.16. A. Cell membrane inhibitors have a difficult time penetrating the highly cross-linked gram-positive cell wall and therefore do not reach the cell membrane.

1.17. D. These media contain nutrients that allow for the growth of certain bacteria only.

1.18. C. Both media provide for the "enrichment" of certain bacteria within the population of bacteria that are capable of growth in it.

1.19. D. *Mesophiles* and *thermophiles* are terms used to characterize the ability of bacteria to grow at various temperatures.

1.20. B. Selective media contain compounds that inhibit the growth of some bacteria.

1.21. D. These media are based on the principle that bacteria have varying biosynthetic capacities and nutritional requirements. Their growth rate constant is irrelevant.

1.22. A. Gram-negative bacteria lack teichoic acids.

1.23. D. Glycocalyx is not part of the cell wall structure.

1.24. C. N-acetylmuramic acid is part of the mucopeptide of both gram-negative and gram-positive cell walls.

1.25. B. Gram-negative bacteria have a Lipid A component in their cell envelope.

23

1.26. B.   Gram-positive bacteria lack an outer membrane structure.

1.27. C.   Both processes involve the oxidation of organic compounds.

1.28. D.   Cyclic photophosphorylation is a biosynthetic process that occurs during photosynthesis.

1.29. A.   Respiration involves the electron transport chain constituents that are located in the bacterial plasma membrane.

1.30. A. O_2 is not used as the final hydrogen acceptor molecule during the oxidation that occurs during fermentation. Respiration requires the reduction of gaseous oxygen.

1.31. C.   Both processes form ATP, which is later used as an energy source.

1.32. B.   Temperate phages synthesize a repressor protein that is necessary for maintenance of the lysogenic replication cycle.

1.33. B.   Temperate phages can form restricted transducing phages capable of transferring only genetic information that is adjacent to their integration site.

1.34. D.   The terms *virulent* and *temperate* are used to describe types of DNA phages that can lyse host cells at some point in their lytic replication cycles. Nonlytic RNA phages attach to male pili.

1.35. C.   Both types of phages can have a lytic cycle that results in host cell lysis.

1.36. A.   Phage typing is dependent on cellular lysis following infection. This process does not always occur with temperate phages.

1.37. B.   Bacteria will multiply faster in complex media than in minimal media. Chemostats and turbidostats maintain cultures at one specific phase of the growth curve.

1.38. D.   The "broad-spectrum" antibiotics, tetracycline, chloramphenicol, the macrolides, and the lincomycins are bacteristatic. They are not effective against stationary phase bacteria and do not give a synergistic effect with the aminoglycosides.

1.39. A.   Endotoxin is attached to the outer membrane by hydrophobic bonds. Some transport proteins for small molecules are found in this structure of gram-negative bacteria. The plasma membrane contains the enzymes involved in oxidative phosphorylation.

1.40. B.   Prokaryotic cells differ from eukaryotic cells in that the former have coupled transcription and translation of m-RNA in the cytoplasm and have a single (haploid) chromosome.

1.41. B.   The biochemical activity of an enzyme may be regulated by binding of allosteric protein or by biosynthetic pathway end product feedback inhibition. Enzyme synthesis may be controlled by inducers or CAP.

1.42. E.   All are correct.

1.43. F.   None are classified as endotoxins.

1.44. F.   *E. coli* heat stable toxin stimulates guanylate cyclase.

1.45. F.   Botulinum exotoxin acts at myoneural junctions.

1.46. F.   *C. perfringens* α-toxin disrupts membranes.

1.47. F.   *Staphylococcus aureus, Clostridium perfringens,* or *Salmonella* cause food poisoning.

2

Properties of Pathogenic Bacteria

Staphylococci

1. General Comments

–cause major problems in hospitals in compromised patients that have debilitating diseases or extensive surgery, or are immunosuppressed or malnourished.
–may be part of normal flora in anterior nares, perineum, gastrointestinal (GI) tract, or skin (carriers).

II. Classification

–*S. aureus* causes most staphylococcal disease.
–*S. epidermidis* causes urinary tract infections and subacute bacterial endocarditis.
–*S. saprophyticus* causes urinary tract infections.

III. Clinical Manifestations

Staphylococcus aureus

–causes disease in almost any organ or tissue.
–hallmark is an abscess that is a central necrotic core of polymorphonuclear leukocytes (PMNs), pus, and bacteria surrounded by an avascular fibroblastic wall of fibrin.
–infections classified as local/contiguous (common) or disseminated (rare).

A. Local/Contiguous Infections

1. Skin

a. folliculitis denotes a superficial infection (mild)
b. boil/furuncle extends into the subcutaneous area (more severe)
c. carbuncle invades deeper tissues, tends to be multiple and contiguous, especially in neck and upper back (most severe)

    d. impetigo involves encrusted pustules on superficial layers of skin, is highly contagious, mostly occurring in preschool children (streptococci also cause impetigo)

  2. Eye: styes and conjunctivitis
  3. Lungs: pneumonia, frequently under 1 year of age, fulminant, may be secondary to influenza
  4. Wounds: trauma or surgery, especially in abdomen

B. Disseminated Infections (Away from Initial Site of Infection)

–two types include bloodstream or lymphatic disseminated infections and toxin-associated infections.

  1. disseminated infections (chronic, difficult to cure)
    –osteomyelitis, especially in diaphysis of long bones.
    –pyoarthritis with permanent cartilage damage, especially in hip.
    –acute bacterial endocarditis in drug abusers or within first 2 months of heart surgery.
  2. toxin-associated infections
    –scalded skin syndrome due to exfoliative toxin that splits epidermis at stratum granulosum, causing denudation of skin; prevalent under age 4 years.
    –toxic shock syndrome due to pyrogenic toxins; mostly in females using tampons; is characterized by high fever with vomiting and diarrhea, collapse of peripheral circulation, hypotensive shock, and scarlitiniform rash with desquamation of skin; low percentage of cases in males.
    –food poisoning due to enterotoxin preformed in food; causes acute vomiting with mild cramps and no fever; occurs 2 to 6 hours after eating contaminated food.
    –enterocolitis following broad-spectrum antibiotics for other bacterial infections; symptoms involve cramps, diarrhea, fever, and dehydration

Staphylococcus epidermidis

–causes urinary tract infections primarily in old age.
–subacute bacterial endocarditis may occur at least 2 months after heart surgery or after GI instrumentation or dental work.

Staphylococcus saprophyticus

–causes urinary tract infections primarily in adolescent females.

IV. Laboratory Diagnosis

A. Identification

–gram-positive cocci in grape-like clusters.
–very hardy organisms relatively resistant to drying, disinfectants, heat, high salt content; may survive for weeks in dried pus.
–growth on blood agar or mannitol salt agar.
–β-type clear hemolysis, catalase positive, mannitol positive (*S. aureus* only), coagulase positive (*S. aureus* only).

B. Clinical Specimens

–if nonsterile site, inoculate blood agar or mannitol salt agar.

–if sterile site (blood, cerebrospinal fluid [CSF]), inoculate thioglycollate or nutrient medium; after growth, then inoculate blood agar or mannitol salt agar.

V. Control

A. Treatment

–Isolate organisms and perform antibiotic sensitivity because of widespread resistance.
–localized infections require oral antibiotic for 10 days.
–disseminated infections require parenteral antibiotic for 4 to 6 weeks.
–drain abscesses, remove foreign body if possible.
–toxic shock syndrome if severe may require IV fluids and elevation of blood pressure.
–food poisoning is self-limiting with 24 hours.

B. Prevention

–hand washing is key prevention in hospital environment.
–hospital problems primarily in operating rooms and nurseries (umbilical stump and groin infections of newborns).
–toxic shock patients should discontinue use of tampons.
–identify and treat carriers, especially in hospital settings.

VI. Virulence Attributes

Enzymes include:

A. coagulase, which enhances fibrin deposition and abscess formation
B. hyaluronidase, which facilitates bacterial spreading
C. penicillinase, which is a plasmid associated β-lactamase responsible for penicillin resistance

Toxins include:

A. exfoliative toxin responsible for scalded-skin syndrome
B. enterotoxins involved in food poisoning and colitis
C. pyrogenic toxins producing toxic shock syndrome
D. Panton-Valentine toxin that damages leukocytes
E. hemolysins that act on cell membranes to lyse cells

–phagocytosis by PMNs is key host defense.
–children with chronic granulomatous disease are very susceptible to repeated staphylococcal infections because of aberrant PMN function.
–high carriage rate that ranges between 20% and 75% of population.
–implanted synthetic devices (sutures, heart valves, venous catheters, CSF shunts) predispose to infection.

Streptococci

I. General Characteristics

–occur as single, paired, or chained gram-positive cocci, depending on environment.
–are facultative anaerobes.
–are heterogeneous.

–attach to epithelial surfaces via lipotechoic acid portion of fimbriae (pili).
–are classified into groups by serology.

II. Classification

–classified into 21 groups (Lancefield groups A-U) through slight differences in specific cell wall carbohydrates.
–also classified according to type of enzymatic hemolysis of red blood cells produced on blood agar plates into:

a. α-hemolysis—incomplete lysis with green pigment surrounding colony
b. β-hemolysis—total lysis and release of hemoglobin; clear area around colony
c. γ-hemolysis—absence of any lysis

III. Group A Streptococci

A. General Characteristics

–prototype termed *Streptococcus pyogenes*.
–contain group A specific carbohydrate and several antigenic proteins, termed M, T, and R antigens, in the cell wall.
–are subdivided into more than 80 types based on antigenic differences in M protein (Example *S. pyogenes*, type 12 streptococcus).
–are sensitive to bacitracin in contrast to other streptococci.
–are catalase negative.
–are β-hemolytic.
–have not become resistant to penicillin.
–can be detected by throat smears on blood agar, latex agglutination tests, or a rapid (10 minute) test using a fluorescein labeled monoclonal antibody.

B. Attributes of Pathogenicity

–possess M proteins, a potent virulence factor found on fimbriae interfering with phagocytosis.
–have a nonantigenic, antiphagocytic hyaluronic acid capsule that promotes invasiveness.
–secrete three serologic types of erythrogenic exotoxins that require lysogenic phage for production and produce the rash in scarlet fever.
–possess two hemolysins:

1. streptolysin S, a protein that is leucocidal and responsible for the β-hemolysis on blood agar plates
2. streptolysis O, which is sensitive to oxygen and also leucocidal

–possess multiple other enzyme systems (e.g., hyaluronidase, streptokinase, streptodornase, nicotinamide adenine dinucleotidase).

C. Clinical Diseases

1. Streptococcal Pharyngitis

–is characterized by sore throat, fever, headache, nausea, cervical adenopathy, leucocytosis.
–can result in complications (e.g., tonsillar abscesses, mastoiditis, septicemia, osteomyelitis, rheumatic fever).

–intense pharyngeal redness, edema of mucous membranes, and a purulent exudate differentiate it from viral infections and infectious mononucleosis.
–treat with penicillin; preferably one effective for 3 weeks.

2. Scarlet Fever

–similar symptoms as streptococcal pharyngitis.
–includes a rash due to erythrogenic toxins.

3. Impetigo (Pyoderma)

–local infection of superficial layers of skin.
–highly communicable in infants.
–may result in nephritis as a complication.

4. Cellulitis-Erysipelas

–initiated by infection through a small break in the skin.
–termed cellulitis if lesion is confined, or erysipelas if lesion spreads.
–can invade subcutaneous tissue and advance rapidly through lymphatics giving rise to a septicemia.

5. Rheumatic Fever

–follows group A streptococcal throat infection in genetically predisposed individuals; however, 20% of patients may show no early signs or symptoms.
–results in a systemic inflammatory process involving connective tissue, heart, joints, and CNS.
–may lead to progressive chronic debilitation.
–may damage heart muscle and valves with mitral stenosis as a lesion hallmark.
–postulated to be caused by streptococcal antigens cross reacting with sarcolemmal muscle and kidney, resulting in a damaging inflammatory process.
–treat with penicillin promptly.
–continue penicillin prophylactically to prevent recurring infections and increased damage.

6. Acute Glomerular Nephritis

–follows generally an earlier skin infection with mainly streptococcus types 2, 4, 12, or 49.
–initiated by deposition of soluble streptococcal antigen-antibody complexes and complement on the glomerular basement membrane; appears as a "lumpy-bumpy" pattern with immunofluorescence.
–induces inflammation via complement incited chemotaxis and PMN infiltration.
–does not generally go on to become chronic glomerular nephritis.

7. Endocarditis

–results from inflammation induced by deposition of any of several bacterial genera on group A streptococcal (or congenitally) damaged heart valves.

IV. Group B Streptococci

A. General Characteristics

–prototype termed *S. agalactiae*.
–occur as part of normal vaginal and oral flora of women.
–occur as five serotypes (Ia, Ib, Ia/c, II, and III) based on antigenic differences in capsular polysaccharides.
–are β-hemolytic.

–synergize with a staphylococcal hemolysin (termed CAMP reaction).
–can be differentiated with group B antiserum, sodium hippurate hydrolysis, and their resistance to bacitracin.

B. Attributes of Pathogenicity

–possess a capsule as the major virulence component.
–anticapsular antibody is protective in presence of competent phagocytic cells and complement.

C. Clinical Diseases

–cause neonatal sepsis in two forms:

1. early onset (birth to 7 days)
   –newborns colonized readily, but only 1/100 become clinically ill
   –infection associated with obstetric complications, premature births, or respiratory distress
   –fatality rate is generally over 50%
2. late onset (7 days to 4 months)
   –characterized by meningitis
   –results often in permanent neurologic damage
   –caused predominantly by type III
   –fatality rate: 15% to 20%

D. Treatment

–vaccine use limited due to poor response of children to polysaccharide antigens.
–penicillin G.

V. Groups C and G Streptococci

–occasionally cause disease in human beings.
–may be increasing in frequency in selected populations.

VI. Group D Streptococci

A. General Characteristics

–prototype termed *S. faecalis*.
–occur as part of the normal intestinal and oral flora in humans and animals.
–produce variable hemolysis but most are α- or γ-hemolytic; *S. faecalis*, however, is β-hemolytic.
–can be differentiated by reactivity with group D antiserum, bacitracin resistance, and growth in 40% bile or pH 9.6.
–can be subdivided into enterococci and nonenterococci.
–enterococci grow in 6.5% NaCl; nonenterococci do not.
–enterococci are inhibited but not killed by penicillin; nonenterococci are killed by penicillin.

B. Attributes of Pathogenicity

–none identified.
–are generally noninvasive opportunists.

C. Clinical Disease

–cause endocarditis, urinary tract infections, and septicemia.

D. Treatment

–test antibiotic sensitivity to determine appropriate treatment.

VII. Viridans Streptococcus

A. General Characteristics

–prototype termed *S. viridans.*
–are unable to be classified by group-specific antigens.
–composed of ten species differentiated by biochemical tests.
–common species are *S. salivarius, S. mutans, S. mitis,* and *S. sanguis.*
–predominate in the normal human oral cavity.
–are α-hemolytic, are uninhibited by optochin, and are not bile soluble.

B. Attributes of Pathogenicity

–none identified.
–are generally noninvasive opportunists.

C. Clinical Diseases

–most frequent cause of bacterial endocarditis.
–important as a major cause of dental caries.

D. Treatment

–generally susceptible to penicillin.
–test antibiotic sensitivity to determine appropriate treatment.

VIII. *Streptococcus Pneumoniae*

A. General Characteristics

–occurs as part of the normal oral flora in 40% to 70% of humans.
–are gram-positive, lancet-shaped diplococci, which are α-hemolytic.
–possesses a group-specific carbohydrate common to all pneumococci, which is precipitated by a C-reactive protein occurring in the plasma of patients undergoing an inflammatory response.
–quantitation of precipitate is an index of extent of inflammation; not an antigen-antibody reaction.
–also possesses a type-specific polysaccharide capsule with over 80 different antigenic types.
–differentiate types by swelling of the capsule in presence of specific antiserum (Quellung reaction).
–injection of relatively large amounts of capsular polysaccharide results in tolerance rather than immunity.
–differentiate from other streptococci by:

1. sensitivity to a quinine derivative, ethyl hydrocuprine, termed optochin
2. sensitivity to bile, which solubilizes pneumococci by increasing an autolytic amidase
3. fermentation of inulin

–differentiation from nonpathogenic *S. viridans* is important as:

1. both are found in the pharynx and sputum
2. both produce α-hemolysis on blood agar
3. both are gram positive and are usually diplococci

B. Attributes of Pathogenicity
  –little evidence for any toxins; most likely produce disease through multiplication.
  –virulence attributed to antiphagocytic capacity of capsule.

C. Clinical Disease
  –cause pneumococcal pneumonia, but rarely as a primary infection.
  –pathogenicity associated with disturbances of normal defense barriers of respiratory tract.
  –most vulnerable are infants, the aged, immunosuppressed, and chronic alcoholics.
  –self-infection can occur by aspiration following slowing of epiglottal reflexes due to chilling, anesthesia, morphine, alcohol, virus infection, or increased pulmonary edema.
  –classical picture includes abrupt onset, fever, chills, chest pain, and productive cough followed by a crisis on days 7 to 10.
  –recovery in 50% to 70% of untreated cases is associated with appearance of anticapsular antibody.
  –two-thirds of deaths occur in first 5 days.
  –identify by culture of sputum from lungs (not saliva) followed by typing via the Quellung reaction.
  –causes otitis media and septicemia in infants older than 1 month.

D. Treatment
  –penicillin or other appropriate antibiotics.
  –an effective vaccine for adults containing at least 23 different type-specific polysaccharides is available.
  –the vaccine is poorly immunogenic in infants.

Neisseria Gonorrhoeae

I. General Comments
  –epidemic; sexually transmitted with the highest incidence in the most sexually active group (ages 15 to 25 years).
  –pyogenic as are staphylococci, streptococci, and hemophilus.

II. Classification
  –*N. gonorrhoeae* differentiated by auxotyping (nutritional requirements) or colonial morphology (types 1 and 2 are virulent; types 3, 4, and 5 are far less virulent).

III. Clinical Manifestations
  –mucous membrane infections that occur predominately in anterior urogenital tract.
  –asymptomatic infections in 20% to 80% of females and 10% of males; these patients transmit the bacteria to consorts, resulting in symptomatic gonorrhea.
  –a number of different types of infection:

  A. Urethritis—thick, yellow, purulent exudate containing bacteria and numerous PMNs; painful and frequent urination; meatus may be erythematous
  B. Complications of urethritis include epididymitis and prostatitis in males and pelvic inflammatory disease in females; repeated infections may cause scarring with subsequent sterility in both sexes
  C. Rectal Infections—painful defecation, discharge, constipation, proctitis; prevalent in gay males

    D. Pharyngitis—mild form mimics viral sore throat; severe form mimics streptococcal sore throat; purulent discharge

    E. Disseminated Infection—bloodstream invasion in which organisms initially localize in skin causing dermatitis (single macular, papular, erythematous lesion); a few days later, organisms spread to the joint causing overt, painful arthritis (hands, wrists, elbows, ankles)

    F. Infant Eye Infection—ophthalmia neonatorum involves severe, bilateral purulent conjunctivitis that may lead to blindness; newborn is infected during passage through birth canal

IV. Laboratory Diagnosis

A. Identification

–gram-negative, intra- and extracellular diplococci and numerous PMNs in purulent exudate in males; in females, because of endocervical localization, not as likely to see a characteristic Gram's stain of organisms.

–culture on Thayer Martin agar in candle jar.

–perform oxidase test (positive) on isolated colonies.

–demonstrate glucose fermentation on isolated colonies.

–newer technique involves immunofluorescence or enzyme-linked immunosorbent assay (ELISA) on direct clinical swab.

B. Clinical Specimens

–on females, always do a genital and a rectal culture.

–if using a speculum or anoscope, do not use a lubricant because it will kill organisms.

–organisms are labile, and specimens should be plated immediately.

–if disseminated gonorrhea, culture blood and synovial fluid; culture of skin lesion is rarely successful.

V. Control

A. Treatment

–aqueous procaine penicillin G with oral probenecid to decrease renal secretion of the antibiotic.

–if gonococci contain penicillinase, treat with spectinomycin.

–if treatment fails, consider concurrent infection with *Chlamydia trachomatis*.

–pelvic inflammatory disease in approximately 50% of cases is severe enough to hospitalize.

–pelvic inflammatory disease after healing, will predispose to repeat episodes of pelvic inflammatory disease caused by other bacteria.

B. Prevention

–treat partners because of contagious nature of gonorrhea.

–to prevent newborn gonococcal conjunctivitis, use topical $AgNO_3$ or tetracycline; antibiotic is better because it will also kill *Chlamydia trachomatis* if present.

–use condom to prevent transmission.

–identify asymptomatic patients by culturing gonococci, and treat with penicillin.

VII. Virulence Attributes

–IgAase that degrades IgA_1; important because in mucosal infections, this antibody probably plays a key early role (IgAase also in hemophilus and streptococcal organisms).

–plasmid that codes for penicillinase production.

–pili are protein surface appendages that mediate gonococcal attachment to mucosal epithelium.

–lipopolysaccharide (LPS) damages mucosal cells.

Neisseria Meningitidis

I. General Comments

–highly fulminant disease prevalent at age 6 months to 2 years; occasional problems in military recruit settings.

–mucous membrane infection of upper respiratory tract.

II. Classification

–16 to 19 different serogroups of *N. meningitidis;* most recent infections in A, B, and C serogroups.

III. Clinical Manifestations

–begins as mild pharyngitis with occasional slight fever.

–in susceptible age group, organisms disseminate to most tissues, especially skin, meninges, joints, eyes, and lungs, resulting in a fulminant meningococcemia that can be fatal in 1 to 5 days.

–initial signs are fever, vomiting, headache, and stiff neck.

–a petechial eruption then develops that progresses from erythematous macules to frank purpura; vasculitic purpura is the hallmark.

–LPS of organisms causes intravascular coagulation, circulatory collapse, and shock.

–fatalities may occur with or without spread to meninges.

–Waterhouse-Friderichsen syndrome describes fulminating meningococcemia with hemorrhage, circulatory failure, and adrenal insufficiency.

–sequelae following recovery involve eighth nerve deafness, CNS damage (learning disabilities and seizures), and severe skin necrosis that may require skin grafting or amputation.

IV. Laboratory Diagnosis

A. Identification

–gram-negative diplococci; Gram's stain CSF and skin lesion aspirates.

–inoculate nutrient broth (blood/CSF) or Thayer Martin agar (skin lesion/pharyngeal swab) and incubate in candle jar.

–test isolated colonies for oxidase (positive).

–perform sugar fermentations on isolated colonies (ferments glucose and maltose; *N. gonorrhoeae* ferments glucose).

–countercurrent immunoelectrophoresis or agglutination reactions to detect capsular polysaccharide in blood/CSF.

B. Clinical Specimens

–organisms are delicate and must be transported to laboratory and processed quickly.

–for Gram's stain of CSF, may have to centrifuge to concentrate organisms.

V. Control

A. Treatment

–the key is early diagnosis and prompt hospitalization based primarily on petechial rash; there are problems with differential diagnosis because rash is similar to the rashes of Rocky Mountain spotted fever, secondary syphilis, rubella, or rubeola.
–treat with high dosage intravenous penicillin; this antibiotic will pass through the inflamed blood-brain barrier.
–provide supportive measures against shock and intravascular coagulation.
–for adrenal insufficiency, corticosteroids may help to stabilize the patient.

B. Prevention

–provide penicillin prophylaxis for exposed young family children; if patient in day-care setting with other young children, also give penicillin; no prophylaxis for older children or adults.
–give rifampin to all family members and to patient to eradicate the carrier state (penicillin will not eradicate carrier state).
–vaccine is capsular polysaccharide from A and C serogroups (B serogroup polysaccharide is poorly immunogenic); biggest problem is vaccine failure in target group aged 6 months to 2 years where most infections occur.

VI. Virulence Attributes

–capsular polysaccharide inhibits phagocytosis.
–LPS causes extensive tissue necrosis, circulatory collapse, intravascular coagulation, and shock.
–IgAase degrades IgA$_1$; probably important because these infections begin on mucosal membranes (streptococci, hemophilus, and neisseria have this enzyme).

Hemophilus

I. General Comments

–severe problems in young children; chronic, less severe infections beyond age 6 years.
–pyogenic as are neisseria, staphylococci, streptococci (pus producing).
–synergy with virus infection that predisposes to severe hemophilus infection; expect incidence to increase during influenza outbreaks.

II. Classification

–*H. influenzae* causes upper respiratory tract infections that may progress to meningitis or epiglottitis; also part of normal throat flora.
–*H. aegyptius* causes bacterial pink eye.
–*H. ducreyi* is a sexually transmitted genital tract infection.

III. Clinical Manifestations

Hemophilus influenzae type b
–begins as mild pharyngitis.

A. Common Infections

–chronic infections producing otitis media, sinusitis, bronchitis in older children and adults.

–in children aged 3 months to 6 years acute bacterial meningitis is rapidly progressive; CNS deficits occur in one-third of cases leading to hydrocephalus, mental retardation, paresis, speech and hearing problems.

B. Rare Infections

–pneumonia in adults over age 50 and young children.
–acute bacterial epiglottitis in young children is rapidly progressive (severe problems within 2 hours, fatal within 24 hours); microabscesses and edema restrict breathing, causing respiratory arrest.
–cellulitis in children.
–pyarthrosis in children.

Hemophilus ducreyi

–causes soft chancre (syphilis is hard chancre) with 1 to 5 umbilicate pustular lesions in genital tract.
–lymphadenitis; draining inguinal node severely swollen and pustular (termed bubo).

Hemophilus aegyptius

–causes highly contagious childhood disease (especially preschool) termed bacterial pink eye.
–purulent conjunctivitis distinguishes it from viral pink eye that exhibits minimal discharge.

IV. Laboratory Diagnosis

A. Identification

Hemophilus influenzae

–gram-negative coccobacillus; Gram's stain of CSF should reveal relatively large number of organisms in meningitis.
–culture CSF, blood, throat on chocolate agar in 10% CO_2 or candle jar.
–grow in presence of X factor (hematin or hemin) and V factor (NAD or NADP).
–coincubation with staphylococci will permit growth in absence of exogenously added X and V factors (termed satellite growth in which staphylococci provide these factors).
–Quellung reaction involves anticapsular antibodies added to the isolated organisms to visualize capsules.
–capsular antigens demonstrated in blood/CSF by latex particle agglutination or by counterimmunoelectrophoresis.

Hemophilus ducreyi

–differentiate from primary stage syphilis by darkfield microscopy (no spiral-shaped organisms) and by serology (nonreactive syphilis tests).
–genital ulcers with bubo formation.

Hemophilus aegyptius

–purulence is key to diagnosis.

B. Clinical Specimens

–*H. influenzae* is very labile; specimens should be rapidly transported to the laboratory and processed immediately.
–CSF may have to be centrifuged to concentrate organisms.

–problem of normal flora: if few organisms are isolated from throat, hemophilus is not the causative agent.

V. Control

A. Treatment

Hemophilus influenzae

–meningitis/epiglottitis are considered medical emergencies requiring immediate hospitalization.
–administer prompt and vigorous antibiotic therapy intravenously for 14 days.
–begin with chloramphenicol and ampicillin or a third generation cephalosporin; after culture of organisms, test antibiotic sensitivity.
–tracheotomy may be required for epiglottitis to prevent respiratory arrest.
–administer rifampin to patient to prevent development of carrier state and subsequent reinfection.

Hemophilus ducreyi

–isolate organism and test antibiotic sensitivity; drain buboes.

Hemophilus aegyptius

–use topical sulfonamides.

B. Prevention

Hemophilus influenzae

–carrier rate varies between 30% and 90%; rifampin should be given to the immediate family as well as to patient.
–warn immediate family, especially if they have other young children, about first symptoms of fever, headache, stiff neck; in day-care center, also warn families of exposed children.
–identify and treat chronic noninvasive hemophilus in adolescents and adults; these patients can infect young children and cause invasive meningitis.
–hemophilus capsular vaccine is available; it does not work well, however, in young children, which is the target population.

Hemophilus ducreyi

–warn patients of sexual transmission; partners should be treated to avoid re-infection; condom will prevent transmission.

VI. Virulence Attributes

–polysaccharide capsule is key to invasive property of *H. influenzae;* it is antiphagocytic and immunosuppressive.
–antibodies to capsule relate to immune status; there is little or no antibody at age 3 months to 1 year, the period of highest rate of meningitis; beyond age 5 years, increasing antibody titers are apparent, and, correspondingly, the incidence decreases.
–anticapsular antibodies promote phagocytosis and killing.
–organisms produce IgAase, which degrades IgA (neisseria and *Streptococcus pneumoniae* also produce this enzyme); this is important since hemophilus is a pathogen of mucous membranes where IgA is one of the key host defense mechanisms.

Bordetella

I. General Comments

–causes whooping cough predominately in children under 1 year of age.

–is part of highly effective DPT vaccine (diphtheria, *pertussis,* tetanus).

–is a classic example of need for continued vaccination; 2,000 to 5,000 cases/year in U.S. despite high rate of vaccination; in countries where vaccine compliance has drastically waned, major outbreaks followed within 5 years.

II. Classification

–*B. pertussis* causes classical whooping cough.

–*B. parapertussis* causes a mild form of whooping cough.

–*B. bronchiseptica* is primarily an animal pathogen that occasionally causes a mild whooping cough in humans.

III. Clinical Manifestations

–is localized only in respiratory tract and does not disseminate.

–paroxysmal cough is hallmark.

–patients exhibit variation in symptoms; in general, the younger the patient, the more severe the disease.

–three distinct clinical stages:

A. catarrhal—mild upper respiratory tract infection with sneezing, slight cough, low fever, runny nose for 1 to 2 weeks

B. paroxysmal—extends to lower respiratory tract with severe cough (5 to 20 forced hacking coughs/20 seconds); little time to breathe creates anoxia and causes vomiting; during this stage tissue damage predisposes to secondary bacterial infections and pneumonia; lasts 1 to 6 weeks

C. convalescent—less severe cough that may persist for several months

–one-third of patients recover without problems; one-third develop neurologic problems; one-third exhibit severe neurologic deficits, i.e., comma, convulsions, blindness, and paralysis probably associated with the anoxia.

IV. Laboratory Diagnosis

–rely primarily on clinical manifestations.

–lymphocytosis (which is unusual for bacterial infections).

A. Identification

–gram-negative coccobacillus.

–plate on Bordet-Gengou agar.

–immunofluorescence either directly on patient specimen or after in vitro growth.

B. Clinical Specimens

–old method is cough plate; newer method is nasopharyngeal swab.

–inoculate agar immediately since organisms rapidly die.

–best chance for positive specimen is during the catarrhal stage; organisms are rarely isolated after 1 to 3 weeks of the paroxysmal stage.

V. Control

A. Treatment

–erythromycin to eradicate *B. pertussis.*

–other antibiotics to prevent secondary bacterial infection; 30% of patients will develop pneumonias due to other organisms.

–supportive measures, i.e., remove secretions, provide oxygen, humidity, watch electrolytes and nutrition (vomiting).

–cough will persist even after erythromycin due to residual toxins.

B. Prevention

–most contagious in catarrhal stage when the highest concentration of organisms is present; isolate patient for 4 to 6 weeks especially from young infants (1 year or less).

–attack rate of 90% in persons that are nonimmune to *B. pertussis*.

–boost contacts under age 4 with vaccine and give prophylactic erythromycin.

–give contacts that are nonimmune, only erythromycin (too late for vaccine).

–start vaccination with the killed whole organism at age 2 months and give three boosters (maternal antibodies do not protect).

–occasional severe side reactions to vaccine involve encephalopathy, CNS abnormalities, convulsions, brain damage in 1 to 300,000 to 5,000,000 recipients; mild side reactions of tenderness and fever routinely occur.

–low rate of severe reactions dictates continued routine use of vaccine.

VI. Virulence Attributes

–pertussis toxin is a single antigen previously termed islet activating protein, lymphocytosis promoting factor, and histamine sensitizing factor; this antigen causes local tissue damage associated with inflammation.

–hemagglutinin pili are responsible for specific attachment to the cilia of epithelium of upper respiratory tract.

–undefined cough toxin is probably neurologically active; cough persists weeks after organisms killed by erythromycin.

Corynebacteria

1. General Comments

–disease is primarily due to elaboration of a very potent exotoxin; diphtheria is the prototype of a toxigenic disease.

–classic example of viral synergy; β-phage has to infect the cornybacterium and bring along the tox$^+$ gene for manufacture of the bacterial exotoxin.

–component of DPT vaccine; primarily a disease of infants and young children with sporadic outbreaks in indigent adults.

II. Classification

–corynebacteria are closely related to mycobacteria and nocardia.

–*C. diphtheriae* causes classic diphtheria.

III. Clinical Manifestations

–diphtheria divided into pharyngeal or cutaneous infections.

A. Pharyngeal Diphtheria

–begins as mild pharyngitis with mild fever and chills.

–spreads up to nasopharynx or down to larynx and trachea.

–edema in cervical nodes, which obstructs breathing, causes characteristic "bull neck" appearance.

–a firmly adherent gray, spreading pseudomembrane is composed of inflammatory necrosis, fibrin, epithelial cells, PMNs, monocytes, and bacteria.
–organisms do not invade but rather elaborate their potent exotoxin that disseminates through the blood to all other tissues, causing hemorrhage and necrosis.
–two major targets of tissue damage are:

1. heart:
   –cardiac dysfunction, myocarditis, and circulatory collapse; occurs within 2 to 3 weeks
2. nerve:
   –motor defects attributed to demyelinization (reversible on healing); cranial nerves are typically affected; and paralysis of throat muscles and polyneuritis, especially of lower extremities, occurs beyond 3 weeks

–infection with tox$^-$ *C. diphtheriae* produces only a mild pharyngitis.
–infection with tox$^+$ *C. diphtheriae* in immune patient produces only a mild pharyngitis.

B. Cutaneous Diphtheria

–presents as a chronic, spreading, grayish skin ulcer.
–most infections occur in southern U.S., especially after insect bite.
–cutaneous form rarely results in toxic damage to heart and nerves; cutaneous form may spread to another patient, causing pharyngeal diphtheria.
–frequent occurrence of Staphylococcus aureus and streptococci as secondary bacterial invaders.

IV. Laboratory Diagnosis

–difficult because of low rate of suspicion.

A. Identification

–gram-positive rods.
–very hardy organisms that can survive in the environment for weeks within sloughed pseudomembrane.
–culture on Löffler's medium and tellurite medium; after 1 day, black or gray colonies appear that microscopically exhibit Chinese letter formation and Babes-Ernst bodies (metachromatic granules).
–demonstrate toxin within isolate:

1. in vitro Elek test involves antitoxin on filter strips that precipitates with toxin elaborated from growing organisms
2. in vivo animals are injected with toxin containing culture filtrate, then antitoxin to demonstrate protection

B. Clinical Specimens

–pseudomembranes "hide" bacteria; debride slightly to loosen, then swab.

V. Control

A. Treatment

–can be rapidly fatal, so urgency in initiating treatment is critical; begin antitoxin without waiting for laboratory findings; once toxin is bound to distant tissues, antitoxin is ineffective.

–give antitoxin IM, or if disease is very severe, give IV.

–antitoxin is from horses, so test for hypersensitivity by injecting diluted antitoxin dermally or dropping antitoxin into eye; within 30 minutes erythema or conjunctivitis indicates hypersensitivity.

–if patient is hypersensitive, continue antitoxin therapy but dilute and have epinephrine ready in case of anaphylaxis.

–give erythromycin.

–monitor cardiac function; may have to use digitalis or antiarrhythmic agent.

–if breathing problems, intubation or tracheotomy may be necessary.

B. Prevention

–vaccine is formalin inactivated diphtheria toxin.

–start vaccination at 2 to 3 months (maternal antibodies are protective), three booster injections thereafter.

–vaccine generates antibodies that neutralize toxin but do not prevent pharyngeal infection or carrier state.

–watch family members of index case, especially infants and young children; boost with vaccine.

–Shick test determines immune status; inject toxin and heated toxin dermally at different sites:

1. at toxin site - if maximum erythema after 5 days, patient is nonimmune; if no erythema after 5 days, patient is immune
2. at heated toxin site - some erythema after 1 to 2 days, which then disappears, patient is hypersensitive to toxin; if no erythema after 1 to 2 days, patient is not hypersensitive to toxin

VI. Virulence Attributes

A. Invasiveness

–organisms can colonize throat (both tox$^+$ and tox$^-$ *C. diphtheriae*).

B. Toxicity (only tox$^+$ *C. diphtheriae*)

–potent polypeptide exotoxin produces almost all histopathology of classical diphtheria; hemorrhage and necrosis within many different tissues.

–two fragments of exotoxin:

1. fragment B is responsible for attachment of toxin to tissues and facilitates transport of fragment A through membranes into cytoplasm
2. fragment A inhibits cellular protein synthesis, eventually killing the cell; this fragment inhibits t-RNA EF-2 elongating factor

Review Test

STAPHYLOCOCCI, STREPTOCOCCI, NEISSERIA, HEMOPHILUS, BORDETELLA, CORYNEBACTERIA

DIRECTIONS: Each of the questions or incomplete statements below is followed by suggested answers or completions. Select the most appropriate one in each case.

A. Staphylococci
B. Neisseria
C. both are correct
D. neither is correct

2.1. Pus production is helpful diagnostic aid.
2.2. Localized infections that may disseminate and cause severe clinical problems.
2.3. Produce urinary tract infections.
2.4. Gram's stain of clinical specimen is helpful diagnostic aid.

A. *Staphylococcus aureus*
B. *Staphylococcus epidermis*
C. both are correct
D. neither is correct

2.5. Associated with subacute bacterial endocarditis 2 months or more after heart surgery.
2.6. Associated with acute bacterial endocarditis within 2 months after heart surgery.

A. *Neisseria gonorrhoeae*
B. *Neisseria meningitidis*
C. both are correct
D. neither is correct

2.7. Ferments glucose and maltose.
2.8. Capsular polysaccharide detection in serum or CSF in disseminated infections.

A. Hemophilus influenzae
B. Neisseria meningitidis
C. both are correct
D. neither is correct

2.9. Meningitis primarily in ages 3 months to 6 years.
2.10. Meningitis primarily in ages 6 months to 2 years.

A. *Bordetella pertussis*
B. *Bordetella bronchisepticus*
C. both are correct
D. neither is correct

2.11. Nonpathogen that does not cause human disease.

A. Corynebacterium diphtheriae
B. Clostridium tetani
C. both are correct
D. neither is correct

2.12. Potent exotoxin is responsible for almost all of the tissue pathology.
2.13. β-prophage inserts the toxin-producing gene into the bacterial genome.

DIRECTIONS: For each of the questions or incomplete statements below, *one* or *more* of the answers or completions given is correct. Choose answer:

A. if only **1, 2,** and **3** are correct
B. if only **1** and **3** are correct
C. if **2** and **4** are correct
D. if only **4** is correct
E. if all are correct

2.14. Clinical manifestations of *Neisseria meningitidis:*
1. rapid progression requires prompt treatment without waiting for laboratory identification of organism.
2. vasculitic purpura is key diagnostic finding.
3. early rash may be confused with Rocky Mountain spotted fever, secondary syphilis, rubella, rubeola.
4. no severe sequelae following recovery.

2.15. Clinical manifestations of *Neisseria gonorrhoeae:*
1. pharyngeal infection is always mild and mimics viral sore throat.
2. dermatitis involves rash over trunk and extremities.
3. ophthalmia neonatorum is always mild and does not lead to blindness.
4. both males and females can be asymptomatic.

2.16. Clinical manifestations of *Hemophilus influenzae:*
1. acute bacterial epiglottitis can be fatal within 24 hours.
2. severe CNS deficits occur in one-third of recovered patients.
3. pus production is typical.
4. synergy with viral infections that predispose to hemophilus; during influenza outbreaks, incidence of *H. influenzae* increases.

2.17. Treatment and prevention of hemophilus infections:
1. for *H. influenzae* meningitis, give rifampin to index case and to family members to eradicate carrier state.
2. bacterial pink eye is treated with intravenous ampicillin.
3. soft chancre is highly contagious; sexual partners should be treated even if asymptomatic.
4. hemophilus capsular vaccine is highly effective in target population of age 3 months to 6 years.

2.18. Laboratory diagnosis of whooping cough:
1. lymphocytosis is prevalent.
2. nasopharyngeal swab is used to isolate organisms from patient.

3. plate clinical specimen on Bordet-Gengou agar.
4. best chance to isolate organism is the paroxysmal stage.

2.19. Treatment and prevention of whooping cough:
1. catarrhal stage is highly contagious.
2. residual cough after antibiotic treatment indicates treatment failure; re-treat with different antibiotic.
3. secondary bacterial infections are frequent.
4. vaccine causes very severe side effects in 1 of 10,000 recipients; do not routinely give vaccine.

2.20. Clinical manifestations of diphtheria:
1. cutaneous infection may spread to another patient and cause pharyngeal diphtheria.
2. bull neck appearance is characteristic.
3. two major problems involve heart and nerve damage due to exotoxin.
4. after infecting pharyngeal area, organisms enter the bloodstream and invade other tissues.

2.21. Laboratory diagnosis of diphtheria:
1. organisms are very delicate and survive poorly outside the host.
2. it is not sufficient just to identify *C. diphtheriae*; it is also important to demonstrate toxin production by the isolate.
3. organisms are impossible to gram stain due to their small size.
4. culture on Löffler's medium or tellurite medium should reveal characteristic Chinese letter formation.

2.22. Group A streptococci
1. are part of the normal intestinal flora.
2. attach to epithelial cells via lipoteichoic acid containing fimbriae.
3. are bile soluble.
4. produce β-hemolysis with streptolysin S.

2.23. Group B streptococci
1. are part of the normal human vaginal flora.
2. are bacitracin resistant.
3. precipitate with *Streptococcus agalactiae* antiserum.
4. cause neonatal sepsis and meningitis.

2.24. A virulence factor(s) for *Streptococcus pneumoniae* is

1. C reactive protein.
2. bacitracin.
3. endotoxic lipopolysaccharide.
4. capsular polysaccharide.

Answers and Explanations

STAPHYLOCOCCI, STREPTOCOCCI, NEISSERIA, HEMOPHILUS, BORDETELLA, CORYNEBACTERIA

2.1. C. Both types of organisms characteristically produce pus.

2.2. C. Both types of organisms disseminate and cause severe manifestions.

2.3. C. Both types of organisms may produce urinary tract infections.

2.4. C. Gram's stain of clinical specimens for both types of organisms is important.

2.5. B. *Staphylococcus epidermis*

2.6. A. *Staphylococcus aureus*

2.7. B. *Neisseria meningitidis* ferments glucose and maltose; *Neisseria gonorrhoeae* ferments only glucose. These are useful diagnostic criteria for differentiating these two organisms.

2.8. B. *Neisseria meningitidis* capsular material is helpful diagnostic tool.

2.9. A. *Hemophilus influenzae.*

2.10. B. *Neisseria meningitidis.*

2.11. D. Neither is correct. *Bordetella bronchiseptica* causes a relatively mild form of whooping cough.

2.12. C. Both organisms elaborate exotoxins that damage tissues, resulting in overt pathology.

2.13. A. *Corynebacterium diphtheriae* has a virus that codes for toxin production.

2.14. A. The only incorrect answer is 4. Severe sequelae are common in treated patients.

2.15. D. The only correct answer is 4. Pharyngeal infection can be severe with manifestations similar to streptococcal sore throat. Dermatitis usually presents as a simple pustule over the inflamed joint. Ophthalmia neonatorium can be extremely severe and cause blindness. Chlamydial eye infections of newborns are relatively mild and do not impair sight.

2.16. E. All four answers are correct.

2.17. B. Bacterial pink eye is relatively mild and can be treated with topical sulfonamides. Hemophilus capsular vaccine is poorly immunogenic in the target population, especially in ages 3 months to 2 years.

2.18. A. Only incorrect answer is 4. Catarrhal stage provides the best chance for isolation. Once patient progresses beyond this stage, it is very difficult to isolate *Bordetella pertussis*.

2.19. B. Residual cough occurs even after successful antibiotic therapy. Vaccine severe side effects occur in a very low percentage of recipients (1 in over 300,000); continue routine immunization of all young children.

2.20. A. The only incorrect answer is 4. Organisms do not invade. The exotoxin enters the circulation and affects other tissues.

2.21. C. Organisms are fairly resistant to environmental influences and can survive for weeks in dried pseudomembranes. Organisms can be routinely gram stained.

2.22. C. Answers 1 and 3 are not correct.

2.23. E. All answers are correct.

2.24. D. Capsular polysaccharide is the virulence determinant.

Nonspore-Forming Anaerobes

1. General Characteristics

–require a reduced oxygen tension (low oxidation-reduction potential) for growth.
–are categorized into:

1. obligate anaerobes, which grow maximally at PO_2 of less than 0.5% to 3%;
2. facultative anaerobes, which can grow either in air (18% oxygen) or in reduced oxygen tension, e.g., *E. coli*;
3. microaerophilic bacteria, which prefer a reduced oxygen tension but will grow on solid media in 10% CO_2.

–are the predominant component of NORMAL human bacterial flora.
–cause infections with both gram-negative and gram-positive rods and cocci.
–most anaerobic infections are polymicrobic, involving more than one genera or species.
–are foul smelling.
–are not communicable or transmissible.
–comprise 99% of total fecal flora (10^{11}/gm of stool in large bowel).

II. Sources of Infection

–outnumber aerobes 1000:1 in gut and 100:1 in mouth.
–found generally proximal to mucosal surfaces; when this barrier is broken anaerobes can invade.
–local anaerobic growth can be achieved when oxygen is used and E_h lowered by aerobic or facultative microorganisms.
–anaerobes can escape into tissues as a result of:

1. gastrointestinal obstruction or surgery
2. diverticulitis
3. bronchial obstruction
4. tumor growth
5. ulceration of intestinal tract by chemotherapeutic agents
6. leukemia-induced gingival bleeding
7. tissue damage during childbirth

III. Laboratory Diagnosis

–transport culture to laboratory immediately in anaerobic transport tubes as oxygen is lethal for these organisms.
–require special media and growth conditions (e.g., thioglycollate broth, agar deeps, Gas Pak anaerobic jars).
–are usually pleomorphic on Gram's stain.
–sputum not suitable for culture due to predominance of nonpathogenic anaerobes.
–obtain specimen by transtracheal puncture or bronchoscopy.

IV. Reasons for Oxygen Intolerance

–may lack superoxide dismutase, catalase, and cytochrome oxidase, enzymes that destroy toxic oxygen metabolic products.

V. Examples of Important Nonspore-Forming Anaerobes

A. *Bacteroides Fragilis*

1. Bacteriology

–are gram-negative rods usually pleomorphic with vacuoles and swelling.
–grow rapidly under anaerobic conditions and are stimulated by bile.

2. Normal Habitat

–most gut anaerobes are *Bacteroides*; 1% of these gut anaerobes are *B. fragilis*.
–found also in female genital tract, but rarely in oral cavity.

3. Attributes of Virulence

–approximately 75% of all *B. fragilis* involved in infection have been encapsulated.
–possess a collagenase and hyaluronidase.
–possess a weak endotoxin but no exotoxin.

4. Infections

–frequent causative agent of brain and gastrointestinal abscesses.
–leading cause of pelvic inflammatory disease (PID).
–cause a cellulitis, particularly in diabetics.

5. Treatment

–debridement and drainage are important in cellulitis.
–most are resistant to tetracyclines and possess a β-lactamase that destroys penicillins and cephalosporins.
–most are sensitive to metronidazole.
–suggested also are clindamycin, kanamycin, and chloramphenicol.

B. *Bacteroides Melaninogenicus*

1. Bacteriology

–are gram-negative, small coccobacilli with occasional long forms.
–have a distinctive black colonial appearance on agar.

2. Normal Habitat

–present mainly as part of the normal flora of the mouth; saliva has 10^9 anaerobes/ml.
–found in low numbers in the gastrointestinal and genitourinary tracts.

3. Attributes of Virulence

–none clearly identified, but possess a collagenase.

4. Infections

–cause of lung abscesses; putrid sputum clue to anaerobic infection.
–cause infections of the female genital tract.

5. Treatment

–similar to *Bacteroides fragilis*.

C. *Fusobacterium Nucleatum*

1. Bacteriology

–are gram-negative, long, slender filaments and fusiform rods.

2. Normal Habitat

–occur in the mouth and occasionally in the stool.
–spreads rapidly.

3. Attributes of Virulence

–none clearly identified.

4. Infections

–synergizes with oral spirochetes resulting in an ulcerating, necrotizing gingivitis (Vincent's angina or trench mouth).
–responsible also for head, neck, and chest infections.

5. Treatment

–sensitive to penicillin and cephalosporin, with penicillin G preferred.

D. Examples of Indigenous Gram-Positive Nonspore-Forming Anaerobes Occasionally Causing Disease

1. Eubacterium

–most commonly isolated from GI tract, frequently with *Bacteroides fragilis*.

2. Propionibacterium

–commonly isolated from skin and GI tract.

3. Peptostreptococcus

–synergizes with *Staphylococcus aureus* and *Bacteroides*.

4. Peptococcus

Clostridia

I. General Comments

–toxemias arising from ingestion of preformed toxins, or from localized infection with release of toxins into the system; most pathology is attributed to well-characterized toxins; tetanus and botulinum toxins are two of the most potent toxins known.
–high fatality rates; *C. perfringens* and *C. tetani* are major problems during wars.
–spores are ubiquitous in soil and play a major role in causing clostridial infections.
–require low oxidation-reduction potential within tissue; clostridia are unable to infect healthy tissues.

II. Classification

–*C. perfringens* causes soft tissue wounds and food poisoning.
–*C. botulinum* causes botulism.
–*C. tetani* causes tetanus.
–*C. difficile* causes gastroenteritis.
–a number of additional pathogenic species contribute to clostridial wound infections.

III. Clinical Manifestations

C. perfringens causes two types of infection:

A. Soft tissue (muscle) wounds following severe trauma (gunshot, car and industrial accidents, compound fractures, septic abortion, hypothermia); organisms elaborate toxins and enzymes to produce gas, edema, impaired circulation; vascular destruction and lactic acid accumulation lowers oxidation-reduction potential with two consequences:

1. anaerobic cellulitis - destruction only of traumatized tissue
2. myonecrosis (gas gangrene) - destruction of traumatized tissue and surrounding healthy tissue; rapidly progressive with shock and renal failure; fatal in 30% of cases.

B. Food poisoning - follows ingestion of contaminated food containing preformed enterotoxin; abdominal pain with severe cramps and diarrhea for 1 day

C. botulinum infects adults and infants:

A. Adult - follows ingestion of preformed toxin in contaminated food; nausea, vomiting, dizziness; cranial palsy, double vision, swallowing and speaking difficulties, muscle weakness; respiratory paralysis and death in 20% of cases; rare cases of wound infection with similar manifestations.
B. Infant - follows ingestion of spores and subsequent germination in GI tract; exotoxin disseminates producing constipation, generalized weakness, and loss of head and limb control (floppy appearance); rarely fatal

C. tetani
–follows minor trauma (lacerations, punctures) or infection of neonatal umbilical cord stump; muscle stiffness, tetanospasms of lockjaw and back arching; short, frequent spasms of voluntary muscles; fatal after several weeks due to exhaustion and respiratory failure

C. difficile
–severe gastroenteritis (also termed pseudomembranous colitis or antibiotic associated colitis) following antibiotic therapy against other bacterial infections; abrupt onset of acute abdominal pain with watery and profuse diarrhea; pseudomembranes contain exudative plaque with underlying necrosis of mucosa; circulatory collapse and death in 30% of cases

IV. Laboratory Diagnosis

A. Identification

–heavy reliance on clinical manifestations since rapid progression precludes laboratory identification.
–gram-positive, short, plump rods; no spores within tissues.
–*C. perfringens*: isolate and demonstrate α-toxin via Nagler reaction in vitro; usually two or three other clostridial species are also involved in soft tissue wounds.
–*C. botulinum*: demonstrate toxin in food, stool, blood, vomitus via injecting mice and protecting via antitoxin.
–*C. tetani*: demonstrate toxin after 24 hours of in vitro culture.
–*C. difficile*: culture stool and demonstrate enterotoxin.

B. Clinical Specimens

–culture immediately under anaerobic conditions.

V. Control

A. Treatment

–hospitalize and begin treatment without waiting for definitive diagnosis.

–antitoxins are effective only if toxins have not yet bound to tissues; so do not delay.

C. perfringens

–anaerobic cellulitis: penicillin plus additional antibiotics to prevent secondary bacterial infections; debride necrotic tissue.
–myonecrosis: penicillin, debride, antitoxin; surgery is likely; do not tightly bandage; hyperbaric oxygen may be helpful.
–food poisoning: self-limiting without treatment.

C. botulinum

–adult: antitoxin, supportive measures for respiratory control, stomach lavage, enemas (no antibiotic).
–infant: good supportive care only, complete recovery without deficits.

C. tetani

–antitoxin, debride tissue, penicillin, tracheotomy to aid breathing, quiet and dark environment to minimize external stimuli that induce spasms.

C. difficile

–vancomycin, watch fluids and electrolytes.

B. Prevention

–biggest problem is widespread occurrence of spores in environment.

C. botulinum

–give antitoxin to all others that ate the contaminated food even if symptoms have not developed.
–heating food 80 to 100°C for 10 minutes inactivates the toxin.
–home canners must use proper sterilization techniques.
–newborns under 1 year should avoid honey.

C. tetani

–vaccine is a component of DPT; boost every 10 years; if major trauma, boost if no previous boost within 5 years.
–in pregnancy, boost to stimulate maternal antibodies that will protect newborn.
–ensure cleanliness of umbilical stump (in third world countries).

VI. Virulence Attributes

C. perfringens

–12 toxins or enzymes; α-toxin is most important in terms of clinical damage (lecithinase that destroys cell membranes).
–enterotoxin associated with food poisoning.

C. botulinum

–exotoxin acts at myoneural junction to produce paralysis of cholinergic nerve fibers with subsequent suppression of acetylcholine release in peripheral nerves.

C. tetani

–exotoxin acts at synaptosomes to obliterate the inhibitory reflex response of nerve fibers and produce uncontrolled impulses; main action is against the brain stem and anterior horns of spinal cord.

C. difficile

–enterotoxin produces GI upset, and cytotoxin kills mucosal cells.

Review Test

NONSPORE-FORMING ANAEROBES AND CLOSTRIDIA

DIRECTIONS: Each of the questions or incomplete statements below is followed by suggested answers or completions. Select the most appropriate one in each case.

A. *Clostridium perfringens*
B. *Clostridium difficile*
C. both are correct
D. neither is correct

2.1. Enterotoxin plays a role in producing clinical manifestations.

2.2. Antibiotic associated colitis following treatment for other bacterial infections.

DIRECTIONS: For each of the questions or incomplete statements below, one or more of the answers or completions given is correct. Choose answer:

      A. if only **1, 2,** and **3** are correct
      B. if only **1** and **3** are correct
      C. if **2** and **4** are correct
      D. if only **4** is correct
      E. if all are correct

2.3. Clinical manifestations of clostridial infections:

1. *C. perfringens* will not infect nontraumatized tissues due to the high Eh of normal tissues.
2. infant botulism results from ingestion of spores with germination in the GI tract.
3. tetanus is fatal after several weeks due to exhaustion and respiratory failure.
4. *C. perfringens* myonecrosis involves spread to and destruction of surrounding healthy tissues.

2.4. Treatment of clostridial infections:

1. anaerobic cellulitis may require amputation or skin grafting.
2. clostridial food poisoning is self-limiting and does not require treatment.
3. infant botulism should be treated with antibiotics to eradicate the organisms.

4. antitoxins must be administered as soon as clostridial infection is suspected; once toxin is bound, the antitoxin is ineffective.

2.5. Prevention of clostridial infections:

1. high incidence of spores in the environment makes prevention extremely difficult.
2. newborns under 1 year should avoid honey, which could be a source of botulinum spores.
3. in third world countries, the incidence of neonatal tetanus can be minimized by ensuring cleanliness of the umbilical stump.
4. there is an effective botulinum vaccine that should be routinely given to newborns to prevent infant botulism.

2.6. Nonspore-forming anaerobes

1. are pleomorphic and can be either gram-negative or gram-positive.

2. are a frequent cause of intra-abdominal infections.
3. occur frequently as mixed infections.
4. include the *Fusobacterium* genus in the oral cavity.

2.7. *Bacteroides fragilis*

1. possess a capsular polysaccharide.
2. is susceptible to penicillin.
3. possesses a β-lactamase.

4. is rarely found in the intestinal tract.

2.8. Effective diagnosis of anaerobic infections requires

1. streaking on eosin-methylene blue agar.
2. rapid transport of culture to the laboratory.
3. assay for superoxide dismutase.
4. special media and growth conditions.

Answers and Explanations

NONSPORE-FORMING ANAEROBES AND CLOSTRIDIA

2.1. C. Both organisms elaborate an enterotoxin.

2.2. B. *C. difficile.*

2.3. E. All answers are correct.

2.4. C Myoneurosis, but not anaerobic cellulitis, may require amputation or skin grafting; anaerobic cellulitis involves only removal of necrotic tissue. Infant botulism does not require antibiotic therapy; the antibiotics rapidly kill the organisms, thereby releasing the toxin and increasing the severity of the manifestions.

2.5. A. The only incorrect answer is 4. A botulinum vaccine that seems to be somewhat effective is available but is only for laboratory workers investigating *C. botulinum.*

2.6. E. All answers are correct.

2.7. B. This organism possesses a capsular polysaccharide and a β-lactamase.

2.8. C. These anaerobes will not grow on eosin-methylene blue (EMB) agar and do not contain superoxide dismutase.

Enterobacteriaceae

I. The Family Enterobacteriaceae (Enterics)

–is composed of hundreds of closely related species and strains inhabiting the large bowel of human beings and animals.

–is difficult to classify since plasmids and DNA are exchanged frequently.

–includes the medically important tribes of *Escherichieae, Serratieae, Salmonelleae, Klebsielleae, Proteae,* and *Yersineae.*

–is composed of gram-negative, nonspore-forming rods, generally facultative anaerobes, that ferment glucose to acid and reduce nitrates to nitrites.

–causes two major disease syndromes: nosocomial infections and gastrointestinal disturbances.

II. Disease Syndromes

A. Nosocomial Infections (Hospital Acquired)

–occur frequently; approximately 2 million/year or 5% to 10% of hospital populations.

–can cause a bacteremia, which frequently results in shock.

–have a high lethality rate (40% to 60%) since many organisms are resistant to antibiotics.

–organisms are normally noninvasive in healthy individuals; however, compromised (cancer, heart, lung disease) and/or immunosuppressed patients are of particular concern.

–outcome depends on extent of preexisting debilitating disease in patient.

B. Gastrointestinal Disturbances

–are usually due to enterotoxins secreted by Enterobacteriaceae.

–need to be differentiated from GI upsets caused by staphylococci, which have a shorter incubation period (6 hours). Enterobacteriaceae generally have a longer incubation period (1 to 2 days).

–due to *Escherichia coli* exhibit three different disease syndromes: enterotoxigenic (traveler's diarrhea); enteropathogenic (infants); and enteroinvasive (dysentery).

III. Differentiation of Genera and Species

–may be based on antigens (serology), biochemical fermentations, and/or growth on differential and selective media.

–is frequently determined by differences in carbohydrate fermentation.

IV. Identification of Pathogen Amongst Normal Flora

–is difficult since the normal flora in intestine vastly outnumbers any pathogen.

–is aided by the knowledge that most pathogens, with the notable exception of *E. coli,* do not ferment lactose.

–is facilitated by the use of differential or selective media catering to the pathogen.

A. Examples of Differential Media

1. Eosin Methylene Blue (EMB)

–differentiates *E. coli* as metallic green colonies, whereas pathogenic, nonlactose fermenting salmonella-shigella are translucent.

–inhibits gram-positive organisms via these aniline dyes.

2. MacConkey's Agar

–differentiates *E. coli* colonies (pink) from the translucent appearance of colonies of pathogens.
–inhibits other organisms by its content of bile salts and crystal violet.

B. Examples of Selective Media

1. Hektoen's Media

–inhibits gram-positive and nonpathogenic organisms
–permits direct plating of feces and growth of pathogen.

2. Salmonella-Shigella Agar

–contains high concentration of bile salts and sodium citrate, which inhibit gram-positive and many gram-negative bacteria, including coliforms.

C. Other Differential Tests

–fluorescent antibody.
–agglutination.
–packaged systems containing multiple carbohydrates and other biochemicals to detect differential fermentations.

V. Antigenic Structure

A. Capsular (K) Antigens

–are generally polysaccharide in nature.
–are exemplified by *Klebsiella pneumoniae* and *Salmonella typhosa*.

B. Flagellar (H) Antigens

–are proteins with antigenically specific and nonspecific phase variations.

C. Fimbriae (Pili)

–are responsible for attachment and colonization of the organism.

D. Somatic (O) Antigens

–are lipopolysaccharides with the terminal sugar(s) the dominant antigen in serologic classification.
–can be classified within the Salmonelleae into a wide variety of serotypes by the Kauffmann-White schema.

VI. Determinants of Pathogenicity

A. Capsule

–suppresses phagocytosis.

B. Enterotoxin (Exotoxin)

–causes a transduction of fluid into the ileum.
–a heat labile and heat stable exotoxin under the genetic control of a transmissible plasmid are produced by *E. coli* (and *Vibrio cholerae*).
–the labile toxin (LT) has two subunits, A and B.

1. Subunit B attaches to the G_{M1} ganglioside at the brush border of epithelial cells of the small intestine and thus facilitates entrance of subunit A.
2. Subunit A activates adenylate cyclase, which increases cyclic AMP, resulting in hypersecretion of water and chloride and inhibition of reabsorption of sodium,

causing electrolyte imbalance. The gut lumen becomes distended with fluid, resulting in hypermotility and diarrhea.

–the heat stable toxin (ST) activates guanylate cyclase in epithelial cells and thus stimulates fluid secretions via cyclic GMP.

C. Endotoxins

–most Enterobacteriaceae possess a lipopolysaccharide complex that is present in the outer membrane, that differs serologically in terminal end sugars (O antigen), but that has a common, toxic, lipid A core causing:

1. Hypotension

–may release endogenous hypotensive agents from platelets and other cells.

2. Fever

–minute amounts (μg) induce interleukin I in human beings.

3. Hemorrhage

–found in adrenals, intestine, heart, kidney.
–produced experimentally in animals in two ways by:

  a. Local Shwartzman reaction - two injections of endotoxin are given: the first intradermally followed in 16 to 36 hours by the second intravenously.
     –results in hemorrhage at id site of first injection. Capillaries plugged with thrombus of platelets, white blood cells (wbc), fibrinoid material.
     –resembles hemorrhage seen in humans during gram-negative bacteremia
  b. Generalized Shwartzman reaction - two injections of endotoxin, both IV, separated by 16 to 36 hours.
     –results in bilateral renal cortical necrosis resembling disseminated intravascular coagulation (DIC)
     –possible mechanism: first injection causes conversion of fibrinogen to fibrin; second injection inhibits phagocytosis of fibrin, and fibrin deposits interrupt circulation in kidney
     –may be produced by only one exposure to endotoxic bacteria during pregnancy or cortisone treatment

4. Adjuvant Action on the Immune Response

–increases antibody response to unrelated antigens.
–stimulates B cells as a mitogen.

5. Increases Resistance to Other Infectious Agents and Tumors

–causes secretion of tumor necrosis factor (TNF).

6. Mediator Release

–causes release of many diverse, physiologically active molecules from macrophages and other cells.

Salmonella

I. General Characteristics

–have a wide host range including human beings, animals, and birds.

–are gram-negative motile rods indistinguishable microscopically from other Enterobacteriaceae.

–are catalogued into over 1,800 serotypes; major concerns are with *Salmonella typhi, S. enteritidis, S. typhimurium, S. paratyphi A, S. schottmülleri,* and *S. choleraesuis,* which cause most human disease.

–do not ferment lactose, but many species are identified by production of acid, gas, and H_2S from glucose.

II. Antigenic Classification

–some possess a capsular (κ) antigen, exemplified by the Vi (for virulence) antigen of *S. typhosa.*

–are grouped via the Kauffman-White schema into over 40 groups based on differences in the oligosaccharide ligands of the O somatic antigens found in the outer membrane.

–assignment of an organism to a particular group is based on the common possession of a major O antigen, which is identified by an Arabic numeral. Example *S. typhi, 9,* 12 and *S. enteritidis* 1, *9,* 12 (and other salmonella) are assigned to Group D as they possess the major antigen, number 9. They also possess other minor antigens, e.g., number 12.

–can be identified further through possession of different flagellar antigens, termed H antigens.

III. Attributes of Virulence

–possess an endotoxin causing diverse toxic manifestations including fever, leukopenia, hemorrhage, hypotension, shock, and disseminated intravascular coagulation.

–some possess an exotoxin (enterotoxin).

–aided by antiphagocytic activity of capsule.

–ability to survive within macrophages by unknown means.

IV. Clinical Diseases

A. Enterocolitis (Also Termed Gastroenteritis or Food Poisoning)

–is presently the most common form of salmonella infection in the USA (approximately 2 million cases/year).

–sources of contamination are multiple: food (most common being poultry and poultry products), human carriers (particularly food handlers), and exotic pets (turtles and snakes).

–most common causative species are *S. typhimurium* and *S. enteritidis,* usually requiring a high infecting dose with a short (8 to 48 hour) incubation period.

–causes a self-limiting illness manifested by fever, nausea, vomiting, and diarrhea.

–is usually characterized by the following pattern of infection:

1. ingestion of organisms in contaminated food;
2. colonization of the ileum and cecum;
3. penetration of the epithelial cells in the mucosa and invasion, resulting in acute inflammation and ulceration;
4. release of prostaglandin by enterotoxins, resulting in activation of adenylcyclase and increased cyclic AMP;
5. increased fluid secretion in large and small bowel.

–blood cultures are usually not positive and agglutination reactions are not helpful.

–requires no specific therapy except to replace fluid loss.

–antibiotic therapy may increase the carrier rate.

B. Septicemic (or Extraintestinal) Disease

–an acute illness most often of nosocomial origin, with abrupt onset and early invasion of the bloodstream.

–characterized by a precipitating incident introducing bacteria (e.g., catheterization, contaminated IV fluids, abdominal or pelvic surgery) followed by a triad of chills, fever, and hypotension.

–wide dissemination of organism may result in local abscesses, osteomyelitis, and endocarditis.

–diagnosis is usually by blood culture, as organisms do not localize in the bowel and stool cultures are often negative.

–may be caused by many species of salmonella as well as other Enterobacteriaceae.

–mortality rate is high (30% to 50%) depending on degree of preexisting debilitation of patient.

–no specific therapy, except maintenance.

C. Enteric Fevers

–produced in the main by *S. typhosa* (typhoid fever) and to a lesser degree by *S. paratyphi* and *S. schottmülleri*, all of which are strict human pathogens.

–occur through the ingestion of food or water, usually contaminated by an unknowing carrier.

–small numbers (e.g., 200 bacteria) are highly infective.

–during an incubation period of 7 to 14 days, the organisms multiply in the small intestine, enter the intestinal lymphatics, and are disseminated via the bloodstream to multiple organs.

–blood cultures then become positive and patient experiences malaise, headache, and the gradual onset of a fever that increases during the day, plateauing at 102° to 105° each day.

–multiplication takes place in the reticuloendothelial system and in the lymphoid tissue of the bowel, producing hyperplasia and necrosis of the lymphoid Peyer's patches.

–a characteristic rash, termed "rose spots" appear in the second to third week in about 90% of patients.

–typical disease lasts 3 to 5 weeks, with the major complications being gastrointestinal hemorrhage and bowel perforation with peritonitis.

–following recovery, 3% of patients become carriers, the organism being retained in the gall bladder and biliary passages. Cholecystectomy may be necessary.

–diagnosis is by isolation of the organism from the blood or stool after 1 to 2 weeks by plating onto differential and/or selective media.

–diagnosis by serology showing rising titers of O antibody is of lesser significance.

–Chloramphenicol is usually the drug of choice, but ampicillin is also effective against most strains.

Shigella

I. General Characteristics

–Gram-negative, facultative anaerobic, nonmotile rods.

–all species of shigella are pathogenic in small numbers for human beings.

–have no known animal reservoir and they are not found in soil or water unless contaminated with human fecal material.

–perpetuation of this disease is due largely to unrecognized clinical cases, convalescent and healthy carriers, less than 1% of which are under the care of a physician.

–disease of poor sanitation, readily transmitted from person to person via food, fingers, feces, and flies.

–all ferment glucose with acid, but rarely gas; only *S. sonnei* ferments lactose.

II. Classification

–classified into four groups based on differences in their somatic O antigens:

1. Group A
   –*Shigella dysenteriae*—rarely found in U.S. unless imported
2. Group B
   –*Shigella flexneri*—common in U.S.
3. Group C
   –*Shigella boydii*—rare in U.S.
4. Group D
   –*Shigella sonnei*—most common cause of disease in U.S.

III. Attributes of Virulence

–all shigellae contain an endotoxic lipopolysaccharide.

–*S. dysenteriae*, type 1 secretes a potent, heat labile protein exotoxin that causes diarrhea and also acts as a neurotoxin.

–possess the capacity to multiply intracellularly, resulting in focal destruction and ulceration.

IV. Clinical Diseases

–cause shigellosis (bacillary dysentery), characterized by acute inflammation of the wall of the large intestine and terminal ileum; bloodstream invasion is rare.

–complications include necrosis of the mucous membrane, ulceration, and bleeding.

–characterized by a sudden onset after a short incubation period (1 to 4 days) of abdominal pain, cramps, diarrhea, and fever.

–stools are liquid and scant and after the first few movements contain mucus, pus, and occasionally blood.

V. Diagnosis

–stool culture of the organism onto differential and selective media.

–serology or blood culture are not helpful.

VI. Treatment

–only *S. dysenteriae* infections require antibiotic therapy.

–resistance to antibiotics has been developing.

–fluid replacement is the most important therapy.

–vaccines under development.

–epidemiologic control by isolation of carriers, disinfection of excrement and proper sewage disposal can be effective.

Campylobacter

I. General Comments

–until recently, only occasional human infections were reported; many of these diarrheas were undiagnosed; better cultivation techniques now indicate a prevalence as high as salmonella and shigella.

II. Classification

–*C. jejuni* causes enteritis (far more prevalent).
–*C. fetus* subspecies *fetus* causes sepsis predominately in debilitated patients with occasional problems during pregnancy and in newborns.

III. Clinical Manifestations

A. *C. jejuni*

–acute enteric illness with invasion and hemorrhagic necrosis of small intestine.
–explosive diarrhea with PMNs, blood, and organisms in stool.
–fever, abdominal cramps, malaise, and headache.
–may confuse with appendicitis or ulcerative colitis.
–if untreated, 20% of patients will have prolonged infection and relapses.

B. *C. fetus* subspecies *fetus*

–acute febrile disease with vascular infection, meningitis, coma, arthritis; frequently fatal.

IV. Laboratory Diagnosis

A. Identification

–spiral- or comma-shaped gram-negative rod.
–stool will exhibit numerous darting, motile organisms plus blood and neutrophils.
–grow in candle jar on selective medium.
–for *C. jejuni*, incubate at 42°C; this organism grows well; other contaminating stool bacteria grow poorly at this temperature.

B. Clinical Specimens

–*C. jejuni* stool.
–*C. fetus* subspecies *fetus* blood.

V. Control

A. Treatment

–*C. jejuni* infection in most cases is self-limiting within 7 days; if more severe or long term, treat with erythromycin.
–*C. fetus* subspecies *fetus* is susceptible to a number of antibiotics.

B. Prevention

–campylobacter infections are associated with domestic and farm animals; outbreaks in past have been attributed to unpasteurized milk and partially cooked poultry.

VI. Virulence Attributes

–an enterotoxin is likely.
–other uncharacterized toxins produce invasion and necrosis of small intestine.

Vibrios

I. General Comments

–cholera is one of the most devastating diseases of humankind (along with plague); pandemics occur that last for a number of years.

–sporadic small outbreaks occur in U.S. along Gulf coast from eating contaminated seafood.

–poor nutrition and debilitation predispose.

–prototype of an enterotoxin-induced diarrhea.

II. Classification

–*V. cholerae* causes classic cholera; most epidemics due to biotypes cholerae and el tor.

–*V. parahaemolyticus* causes relatively mild gastroenteritis.

III. Clinical Manifestations

–vibrio infections result from ingesting contaminated water or food.

A. *V. cholerae*

–abrupt onset of intense vomiting and diarrhea is key finding.

–copius fluid loss (15 to 20 liters/day) leads to rapid metabolic acidosis and hypovolemic shock.

–eyes and cheeks are sunken, skin turgor is diminished.

–remission or death after 2 or 3 days.

B. *V. parahaemolyticus*

–causes gastroenteritis and diarrhea.

–remission after 3 days.

IV. Laboratory Diagnosis

A. Identification

–clinical manifestations combined with a history of living in, or a recent visit to, endemic area.

–gram-negative short-curved rod ("comma shaped") in stool specimen.

–fluorescent antibody test on stool specimen.

–if necessary, can culture on selective medium, then fluorescent antibody or slide agglutination test on organisms from isolated colonies.

B. Clinical Specimens

–stool specimens are clear and watery (rice water stools).

–organisms are very sensitive to acid pH, therefore, culture stool without delay.

V. Control

A. Treatment

V. cholerae

–key is prompt replacement of fluids and electrolytes; patient should appear healthier within 1 to 3 hours.

–initially use IV route; as patient responds, switch to oral route.

–fluid electrolyte therapy reduces fatality rate from 60% to 1%.

–give tetracycline to prevent patient from infecting others; if no antibiotics, patient recovers but will shed organisms for 1 year.

V. parahaemolyticus

–if mild, no treatment is required.

–if more severe, may have to hospitalize to balance fluids and electrolytes.

B. Prevention

V. cholerae

–adequate sewage disposal and water purification are keys.

–hospitalize patients since rice water stools are highly contagious.

–identify and treat carriers with tetracycline.

–currently endemic in India and Bangladesh.

V. parahaemolyticus

–refrain from swimming in contaminated estuaries or eating contaminated seafood (proper cooking will kill these organisms).

VI. Virulence Attributes

–almost all pathology is attributed to the protein enterotoxin termed choleragen.

–enterotoxin has A fragment (toxic action) and B fragment (attachment to cells).

–enterotoxin specifically attaches to epithelial cells of microvilli at brush borders of small intestine.

–enterotoxin stimulates adenyl cyclase to overproduce cyclic AMP, which upsets the fluid/electrolyte balance; there is hypersecretion of chloride and bicarbonate.

Opportunistic Gram-Negative Bacteria

I. General Characteristics

–composed of Enterobacteriaceae found as part of the normal flora.

–are generally noninvasive unless host is compromised in some way.

–usually nosocomial infections.

–often produce similar disease entities including urinary tract and wound infections, pneumonia, meningitis, septicemia, and gastrointestinal disorders.

II. Representative Organisms

Escherichia Coli

–cause:

1. ascending urinary tract infections, which may progress from urethritis to pylonephritis
2. gastroenteritis
3. pneumonia (by aspiration)
4. neonatal meningitis
5. septicemia

Klebsiella

–cause pneumonia and urinary tract infections.

–virulence related to large capsule.

–resistant to ampicillin, but sensitive to cephalosporins.

Serratia

–cause bacteremia, pneumonia, endocarditis.

–pigmented (pink or salmon-red).

Proteus

–cause bacteremia and pneumonia.
–highly motile.
–produce urease, raising pH of urine.

Review Test

SALMONELLA, SHIGELLA, CAMPYLOBACTER, VIBRIO, ESCHERICHIA, KLEBSIELLA, ENTEROBACTER, SERRATIA, PROTEUS

DIRECTIONS: Each of the questions or incomplete statements below is followed by suggested answers or completions. Select the most appropriate one in each case.

A. *Campylobacter jejuni*
B. *Campylobacter fetus* subspecies *fetus*
C. both are correct
D. neither is correct

2.1. Sepsis is a problem in debilitated patients.
2.2. Human infections are rarely encountered relative to salmonella and shigella infections.

A. *Vibrio cholerae*
B. *Vibrio parahemolyticus*
C. both are correct
D. neither is correct

2.3. Nonpathogen that does not cause human disease.

A. Salmonella
B. Shigella
C. Both of the above
D. neither of the above

2.4. Septicemia.
2.5. A large animal reservoir.
2.6. The mucosal epithelial cells are invaded.
2.7. Translucent colony growth on MacConkey agar.

DIRECTIONS: For each of the questions or incomplete statements below, *one* or *more* of the answers or completions given is correct. Choose answer:

- A. if only **1, 2,** and **3** are correct
- B. if only **1** and **3** are correct
- C. if **2** and **4** are correct
- D. if only **4** is correct
- E. if all are correct

2.8. *Vibrio cholerae* infections:
1. rapidly progressive and may be fatal within 2 days of manifestations.
2. prototype of enterotoxin disease.
3. one of the two most devastating diseases of humankind.
4. metabolic acidosis and hypovolemic shock develop.

2.9. Treatment and prevention of cholera:

1. once treatment is started, recovery is slow and patient remains sick for a few weeks.
2. "rice water stools" are highly contagious and provide the primary mechanism of spread.
3. effective vaccine is available; it should be routinely given to persons traveling to endemic area.
4. tetracycline is administered primarily to stop transmission of the organisms to exposed people.

2.10. Enterobacteriaceae

1. are frequent causes of hospital-acquired infections.
2. are usually noninvasive and part of the normal intestinal flora.

3. can have both endotoxins and exotoxins.
4. ferment glucose.

2.11. Endotoxic activity is associated with

1. the O specific polysaccharide side chain of gram-negative bacteria.
2. lipid A.
3. flagella from gram-negative bacteria.
4. lipopolysaccharides from gram-negative bacteria.

2.12. Bacterial shock

1. is usually a hospital-induced syndrome.
2. can occur after catheterization.
3. mortality is dependent on admission prognosis.
4. is generally preceded by a fever spike.

Answers and Explanations

SALMONELLA, SHIGELLA, CAMPYLOBACTER, VIBRIOS, ESCHERICHIA, KLEBSIELLA, ENTEROBACTER, SERRATIA, PROTEUS

2.1. B. *C. fetus* subspecies *fetus*.

2.2. D. *C. jejuni* infections are at least as common as salmonella and shigella infections.

2.3. D. *V. parahaemolyticus* causes a relatively mild form of gastroenteritis.

2.4. A. Salmonella.

2.5. A. Salmonella.

2.6. B. Shigella.

2.7. C. Both are correct.

2.8. E. All answers are correct.

2.9. C. As soon as fluids and electrolytes are administered, expect to see a very rapid recovery of patients within hours. No vaccine is available; the cholergen toxin is so potent and fast acting that a vaccine-induced immunity is unhelpful.

2.10. E. All answers are correct.

2.11. C. Lipid A exhibits endotoxin activity; gram-negative organisms possess endotoxic LPS activity.

2.12. E. All answers are correct.

Legionella

I. General Characteristics

Legionella pneumophila

–is a gram-negative rod-shaped bacterium that may form longer filaments.
–is stained well only with a Dieterle silver stain.
–is a facultative intracellular parasite.
–has a high density of branched fatty acids.
–is catalase positive and most strains are weakly oxidase positive.
–hydrolyzes hippurate. (Other species of *Legionella* do not.)
–is a stream bacterium that contaminates air-conditioning cooling towers.

II. Classification

Legionella pneumophila

–is classed in a new family and genus of aquatic organisms.
–has at least ten serotypes.

III. Attributes of Virulence

Legionella pneumophila

–grows intracellularly and fails to activate the alternate complement pathway.
–produces cytotoxin, a small peptide, interfering with oxygen-dependent processes of phagocytosis.
–produces β-lactamases to inactivate cephalosporins and penicillins.
–has an endotoxin.
–produces a variety of hemolysins and proteolytic enzymes.

IV. Clinical Disease

A. Pneumonia (Legionnaires' Disease)

–organism enters the body by inhalation. (Smokers are at higher risk.)
–is an acute fibrinopurulent pneumonia, usually unilateral with headache, confusion, fever and chills, and GI tract symptoms of pain, vomiting, and diarrhea.
–peaks in July to October with clusters of cases usually resulting from air-conditioning units or streams.
–is responsible for clusters of hospital-acquired pneumonias.
–is not contagious.
–is diagnosed by:

a. direct fluorescent antibody staining of specimens.
b. culture on special media.
c. demonstrating a rise in antibody titer.

–is isolated only from the lung although GI tract and CNS symptoms occur with the pneumonia.
–is treated with erythromycin with or without rifampin.
–is prevented by the decontamination of cooling towers.

B. Pontiac Fever

–is a mild respiratory disease with pleuritis (no pneumonia).
–is also caused by *Legionella pneumophila*.

Pseudomonas

I. General Characteristics

–is a small, polarly flagellated, gram-negative rod with pili.

–is a nonfermenter of lactose.

–may form a mucoid polysaccharide capsule, particularly in lung colonization of cystic fibrosis patients.

–often produces pigments that may be clinically useful: fluorescein, a greenish fluorescent pigment; pyocyanin, a blue-green pigment.

–blue-green pus is a classic sign of *Pseudomonas aeruginosa* burn infection.

II. Classification

–*Pseudomonas aeruginosa* is the most common causative agent and the type species.

–can be typed using the envelope lipopolysaccharide (LPS) or the pigment pyocyanin.

–has numerous other species, some of which are *Ps. cepacia, Ps. maltophilia,, Ps. mallei, Ps. pseudomallei.*

III. Attributes of Virulence

–invasive factors include pili, which attach to damaged basement membrane of cells; capsular polysaccharide, which increases adherence to tissues but does not decrease phagocytosis; a hemolysin with phospholipase activity; collagenase and elastase; and flagella to aid in motility.

–virulence factors include: (1) LPS, similar to that found in the Enterobacteriaceae; (2) Exotoxin A, an ADP-ribose transferase similar to diphtheria toxin, which halts protein synthesis and causes necrosis in the liver; (3) Exotoxin S, which also is an ADP-ribose transferase capable of inhibiting eukaryotic protein synthesis.

IV. Clinical Disease

–is predisposed by neutropenia, prolonged antibiotic use, severe burn, cystic fibrosis, trauma, metabolic disease, or inhalation of large numbers of organisms such as occasionally happens in contaminated respiratory therapy fluids.

–is generally accompanied by symptoms of endotoxin shock, disseminated intravascular coagulation, and adult respiratory distress syndrome.

–frequently begins with overgrowth of *Pseudomonas* in the intestine of antibiotic-treated, hospitalized patients.

–may occur in any site of the body but most commonly in the urinary tract.

A. Cellulitis

–occurs in patients with burns, wounds, or in neutropenia.

–is indicated by blue-green pus and a grape-like sweet smell.

–may be highly necrotic.

B. Pneumonia

–colonization of lungs with mucoid strains occurs in cystic fibrosis.

–may occur in patients exposed to high levels of *Pseudomonas* in contaminated inhalation therapy equipment.

–often results in mental confusion, gram-negative septic shock, and cyanosis of increasing severity.

C. Septicemia

–is a result of hematogenous spread from local lesions.

–results in ecthyma gangrenosum when dermal veins and tissues are invaded.

D. Other Opportunistic *Pseudomonas* Infections Include

–urinary tract infections, which are generally chronic and occur in the elderly.

–endocarditis in drug addicts.

E. Meliodosis

–initial infection may go undiagnosed and be followed years later by fulminant septicemia.

–is caused by *Ps. pseudomallei.*

V. Diagnosis

–is most commonly made by clinical suspicion (grape odor, blue-green pus or ecthyma gangrenosum of gram-negative septicemia) and confirmed by culture.

–most strains show β-hemolysis on blood agar with pigment production.

VI. Treatment

–is difficult because of frequent resistance to antibiotics.

–requires combination therapy (carbenicillin or ticarcillin plus aminoglycoside) until drug susceptibilities are determined.

VII. Prevention

–is difficult but incidence can be reduced (1) in burn units by careful sanitization of drains and aerators, by a ban on plants and raw vegetable foods, etc; (2) in respiratory therapy by sterilization of equipment; (3) and eventually in some susceptible patient populations by vaccination.

Chlamydia

I. General Comments

–obligate intracellular parasites because they cannot synthesize ATP; have a rigid cell wall; resemble gram-negative bacteria.

–exist in two forms: (1) *elementary body*, which is infectious, and (2) *reticular body*, which is the intracellular reproductive form.

II. Classification

–*C. psittaci* is a disease of birds (ornithosis), particularly psittacine birds (parrots), that causes respiratory tract infection of humans.

–*C. trachomatis* differentiated into 15 serotypes: (1) A, B, and C cause trachoma; (2) D–K cause genital infections; and (3) L causes lymphogranuloma venereum.

III. Clinical Manifestations

A. *C. psittaci*

–cause a wide spectrum of human respiratory disease ranging from subclinical infections to a fatal pneumonia.

–characterized by a sudden onset that initially resembles influenza.

B. *C. trachomatis*

1. Subtypes A, B, and C

–cause a chronic keratoconjunctivitis *(trachoma)* that can progress to conjuctival and corneal scarring and blindness.

–frequently there is a concomitant secondary bacterial infection.

2. Subtypes D–K

–produce a sexually transmitted disease that may also have an associated inclusion conjunctivitis.
–are a prominent cause of nongonococcal urethritis in males, and urethritis, cervicitis, scalpingitis, and *pelvic inflammatory disease (PID)* in females.
–produce a relatively high incidence of asymptomatic or relatively inapparent infections.
–can produce a usually self-limiting *inclusion conjunctivitis in infants* delivered through an infected birth canal.

3. Subtype L

–produces a sexually transmitted disease called *lymphogranuloma venereum,* which is characterized by a suppurative inguinal adenitis.
–lymphadenitis may progress to lymphatic obstruction and rectal strictures if diseased is untreated.

IV. Lab Diagnosis

A. Identification

–appropriate specimens (tissue scrapings or swabs) are (1) inoculated onto tissue culture cells (frequently McCoy cells), which allow infection by the organisms, or (2) stained with specific fluorescent antibodies against chlamydial antigens.
–inoculated cell cultures are stained for typical intracytoplasmic inclusion: *C. psittaci*—diffuse inclusions that lack glycogen and do not stain with iodine; *C. trachomatis*—compact inclusions that contain glycogen and stain with iodine.

B. Clinical Specimens

–*C. psittaci*—inclusions in conjunctival epithelial cell scrapings; inoculation of conjunctival scrapings into cell culture; and serology for type-specific antibodies.
–*C. trachoma subtype L*—pus and biopsy material inoculated into cell cultures, which are then examined for typical inclusions or chlamydial antigens detected by fluorescent antibodies; serologic testing for complement-fixing antibodies; *Frei test*—skin test available outside U.S.; no licensed antigens available in the U.S.

V. Control

A. Treatment

–*C. psittaci*—tetracycline.
–*C. trachomatis*—sulfonamides and tetracycline.

B. Prevention

–*C. psittaci*—improvement of hygienic standards and antibiotics.
–*C. trachomatis*—diagnosis of infected mother in neonatal infections; and standard control measures, i.e., use of condoms to prevent sexual transmission.

VI. Virulence Attributes

–some toxic effects of antigens that kill the host cell.
–outer cell wall resembles gram-negative bacteria, but N-acetylmuramic acid is not present.
–reticular body divides by binary fission in an intracellular vacuole.

Rickettsia

I. General Comments
–cause a zoonotic disease in which arthropods are vectors of human disease.
–induce variable clinical manifestations (benign and self-limiting to highly fulminant causing many deaths).
–are intracellular bacteria with a specific predilection for endothelial cells of capillaries.

II. Classification
–family *Rickettsiaceae*.
–genera:

Rickettsia (typhus fevers, spotted fevers, scrub typhus)
Coxiella (Q fever)
Rochalimaea (trench fever)

III. Clinical Manifestations
–multisystem diseases of endothelial cells resulting in hyperplasia, thrombus formation, inhibited blood supply, angiitis, peripheral vasculitis.
–*general rickettsial manifestations* involve *abrupt onset of* high fever, chills, headache (severe, frontal, unremitting), myalgias; a few days later hemorrhagic rash, stupor, delirium, and shock develop.
–epidemic typhus (*R. prowazeckii* - carried by lice): initial trunk rash, then spreads to extremities, gangrene of peripheral limbs, renal and heart failure, shock, high fatality rate; Brill-Zinsser disease is reactivated, mild form of epidemic typhus that follows years after initial infection.
–endemic typhus (*R. typhi* - carried by fleas, also called murine typhus): initial trunk rash, then spreads to extremities; manifestations far less severe than epidemic typhus.
–Rocky Mountain spotted fever (*R. rickettsii* - carried by ticks): initial rash on extremities, then spreads to trunk; can be high fatality rate.
–rickettsial pox (*R. akari* - carried by mites): rash similar to chickenpox, regional adenopathy; eschars are indurated lesions with necrosis; benign course.
–Q fever (*C. burnetii* - transmitted by inhalation of infected dust or by ticks): no rash, pneumonitis that resembles atypical pneumonia, hepatitis; subacute bacterial endocarditis may occur months to years later.

IV. Laboratory Diagnosis
A. Identification
–heavy reliance on clinical manifestations, especially rash and abrupt onset of fever, headache, chills with recent exposure to arthropods.
–acute versus convalescent sera using agglutination reactions and cross-reacting proteus antigens, immunofluorescence reactions, and complement fixation tests.

B. Clinical Specimens
–blood for antibody detection.

V. Control
A. Treatment
–tetracycline is drug of choice.

–chloramphenicol should be used if no time to differentiate between Rocky Mountain spotted fever and *Neisseria meningitidis.*
–if suspect Rocky Mountain spotted fever, initiate treatment immediately.
–instruct patients to continue prescribed regimen of antibiotics; after rash disappears, patients tend to discontinue treatment.
–expect treatment failures.

B. Prevention

–avoid vectors.
–endemic areas in U.S.:

Rocky Mountain spotted fever mostly in Appalachian states;
endemic typhus in southwestern states;
rickettsialpox in northeastern states in urban areas;
Q fever in California, Texas, Illinois farm settings.

VI. Virulence Attributes

–intracellular residence within endothelial cells of vascular system.
–cross-reacting antigens with proteus species (serologic testing).

Mycoplasma

1. General Comments

–smallest and simplest of the procaryotes that are self-replicating.
–lack a cell wall (unique for bacteria); not to be confused with bacterial L-forms, which lack a cell wall in presence of antibiotics; L-forms revert back to cell wall forms after antibiotic is removed.
–requirement for cholesterol (unique for bacteria).

II. Classification

–*Mycoplasma pneumoniae* (also termed PPLO or Eaton agent) causes respiratory tract infections.
–*Mycoplasma hominis* and *Ureaplasma urealyticum* appear to be involved in genital tract infections.

III. Clinical Manifestations

–mycoplasma are mucous membrane pathogens that do not invade other tissues.

M. pneumoniae

–primary atypical pneumonia sometimes referred to as walking pneumonia.
–slow onset of fever, throbbing headache, malaise, and nonproductive cough; over a period of several weeks interstitial or bronchopneumonic pneumonia develops; x-rays reveal segmental lobar pneumonia.
–highest incidence at 5 to 15 years.
–accounts for one-third of all teenage pneumonias.

U. urealyticum

–urethritis, prostatitis, and pelvic inflammatory disease are attributable to this organism.

M. hominis

–pelvic inflammatory disease, postpartum fever, and postabortal fever are attributable to this organism.

IV. Laboratory Diagnosis

A. Identification

–in vitro culture is not routinely attempted; rely on clinical presentation and serology.

–culture on PPLO agar; may not form visible colonies for 2 to 3 weeks; colonies exhibit a characteristic "fried egg" appearance.

–Giemsa stain on cultured organisms reveals small pleomorphic bacteria.

B. Clinical Specimens

–acute versus convalescent sera; complement fixation test and cold agglutinin reaction.

–nasopharyngeal secretions.

V. Control

A. Treatment

–tetracycline or erythromycin over prolonged period will help to resolve manifestations; *M. pneumoniae* will still be shed by these treated patients.

–give tetracycline or erythromycin for *M. hominis* or *U. urealyticum*.

B. Prevention

–reinfections with *M. pneumoniae* are common.

–*U. urealyticum* is sexually transmitted so treat partners.

–condom will prevent *U. urealyticum* transmission.

VI. Virulence Attributes

–LPS, which is quite different from that of gram-negative bacteria.

–glycolipid fraction may play a role in autoimmune-like reactions.

–H_2O_2 release may damage epithelial cells.

Review Test

LEGIONELLA, PSEUDOMONAS, CHLAMYDIA, RICKETTSIA, MYCOPLASMA

DIRECTIONS: Identify the lettered phrase most closely related to the numbered statement.

A. *Rickettsia prowazeckii*
B. *Rickettsia typhi*
C. *Rickettsia rickettsii*
D. *Coxiella burnetii*
E. all are correct
F. none are correct

2.1. May be spread via inhalation of infected dust.

2.2. Characterized by abrupt onset of fever, chills, and unremitting headache.

2.3. Initial rash appears on extremities, then spreads to trunk.

2.4. Zoonosis spread to humans by arthropod vectors.

2.5. Obligate intracellular bacteria with predilection for multiplication within capillary endothelial cells.

A. *Mycoplasma pneumoniae*
B. bacterial L-forms
C. both are correct
D. neither is correct

2.6. Lack bacterial cell wall.

DIRECTIONS: For each of the questions or incomplete statements below, *one* or *more* of the answers or completions given is correct. Choose answer:

    A. if only **1, 2,** and **3** are correct
    B. if only **1** and **3** are correct
    C. if **2** and **4** are correct
    D. if only **4** is correct
    E. if all are correct

2.7. *Mycoplasma pneumoniae*:

1. Eaton agent causes primary atypical pneumonia.
2. Surface mucous membrane infections that do not disseminate to other tissues.
3. Slow-growing organism that requires 1 to 3 weeks to culture in the laboratory.
4. Serologic tests involve complement fixation or cold agglutination.

2.8. *Pseudomonas*

1. gram-positive.

2. ferments lactose.
3. belongs to the family Enterobacteriaceae.
4. often has polysaccharide slime layer.

2.9. Pseudomonas virulence and invasive factors include

1. Exotoxin A.
2. Endotoxin.
3. Exotoxin S.
4. Phospholipase-hemolysin.

76

2.10. *Legionella*

1. causative agent of a fibrinopurulent pneumoniae.
2. weakly gram-negative.
3. faculative intracellular parasite.
4. associated with water.

2.11. Legionnaires' disease

1. is contagious.
2. often occurs in epidemics.
3. is a granulomatous disease.
4. may present with headache, malaise, and myalgia.

Answers and Explanations

LEGIONELLA, PSEUDOMONAS, CHLAMYDIA, RICKETTSIA, MYCOPLASMA

2.1. D. *C. burnetii* is the only rickettsial disease that can be spread via dust particles. All other rickettsia rapidly die in the environment.

2.2. E. Rickettsial diseases in general exhibit these three manifestations.

2.3. C. Rocky Mountain spotted fever rash characteristically appears initially on hands and feet.

2.4. E. All answers are correct. *C. burnetii* is usually transmitted by dust particles, but can be transmitted by ticks.

2.5. E. All answers are correct.

2.6. Both are correct. The L-forms can revert back to the cell wall containing parent form on removal of the cell wall inhibitory antibiotics.

2.7. E. All four answers are correct.

2.8. D. *Pseudomonas* is a gram-negative, nonfermentor belonging to the family Pseudomonadaceae. It is capable of forming a capsular polysaccharide.

2.9. E. The two exotoxins stop protein synthesis in host cells; the phopholipase hemolysin adds to the invasiveness and probably damages lung surfactant; endotoxin causes shock, thrombocytopenia, leukopenia, and DIC.

2.10. E.

2.11. C. Epidemics of Legionnaires' disease occur through common point exposure; central nervous system symptoms and GI tract symptoms occur despite the fact that the organisms are found only in the respiratory system.

Listeria Monocytogenes

I. General Comments
–zoonosis in vertebrate animals (birds, fish, mammals).
–intracellular parasite of mononuclear phagocytes; frequently used as model for immune responses to intracellular organisms.

II. Classification
–*L. monocytogenes* divided into four serogroups.

III. Clinical Manifestations
–protean manifestations of abscesses and granulomas in many different tissues; high mortality rates of 30% to 80%.
–PMN leukocytosis with eventual development of monocytosis.
–organisms enter through ingestion and subsequently disseminate from the GI tract.
–most listeriosis occurs in neonates or in older compromised adults; incidence is increasing because there are more compromised hosts.
–neonatal disease arises from mild genital tract infection with flu-like symptoms during last trimester of pregnancy; two different syndromes:

A. In Utero Infection - perinatal septicemia resulting in abortion, stillbirth, or death a few days after birth; widespread granulomas in most tissues
B. Infection During Passage Through Birth Canal - purulent meningitis that kills newborn in 1 to 4 weeks

–adult disease involves a purulent meningitis in compromised hosts (corticosteroid or radiation therapy, renal transplants, lymphomas, alcoholics, collagen-vascular diseases).

IV. Laboratory Diagnosis
A. Identification
–gram-positive coccobacillus (intra- and extracellular).
–culture on blood agar and demonstrate hemolysis.
–characteristic "tumbling motility" after in vitro growth.
–Anton test involves inoculation of rabbit eye that produces purulent conjunctivitis with keratitis.
–frequent problem in differentiating from streptococcus or corynebacteria.

B. Clinical Specimens
–CSF, blood, amniotic fluid, genital tract of mother.
–cold enrichment technique involves storage of specimen at 4°C for 3 months with periodic weekly or monthly culture at 37°C.

V. Control
A. Treatment
–prognosis is poor with high fatality rates in neonates.
–ampicillin intravenously.

B. Prevention
–major problem is animal reservoirs.
–milk pasteurization kills organisms.

–recognize infections during pregnancy and treat immediately (difficult because of mild symptoms).

VI. Virulence Attributes

–intracellular residence within mononuclear phagocytes.

–hemolysin disrupts vascular membranes within phagocytes enabling organisms to grow intracellularly.

Yersinia

I. General Comments

–zoonosis: plague occurs in wild rodents especially rats (flea spread); yersiniosis occurs in mammals and birds (food and water spread).

–plague is one of the most devastating diseases of mankind; "black death" is rapidly progressive and very contagious with high fatality rates.

II. Classification

–*Y. pestis* causes plague.

–*Y. pseudotuberculosis* and *Y. enterocolitica* cause yersiniosis.

III. Clinical Manifestations

Y. pestis

–after flea bite, organisms migrate to regional lymph nodes, enter the bloodstream, then seed the spleen, liver, and lungs.

–sudden onset of fever, conjunctivitis, and regional bubo are hallmarks of early disease; malaise, nausea, limb and back pain; rapid vascular collapse with disseminated intravascular coagulation and purpuric skin lesions over entire body ("black plague") are characteristic.

–bubonic plague results from infected flea bite; death occurs in 3 to 5 days.

–pneumonic plague results from human to human transmission via respiratory droplets; death occurs in 2 days.

Yersiniosis

–lesions develop in wall of small intestine producing severe GI disease with mesenteric lymphadenitis.

–patient presents with fever, diarrhea, and severe abdominal pain; extensive loss of fluids and blood may occur.

–two clinical courses:

A. Acute lymphadenitis, terminal ileitis, or enterocolitis; this is the most common of the two courses and occurs primarily in young children; it mimics acute appendicitis

B. Septicemia with abscesses in various organs; this disease occurs predominantly in debilitated patients

IV. Laboratory Diagnosis

A. Identification

–gram-negative coccobacillus showing bipolar staining with Wayson's reagent.

–Gram's stain of sputum and infected node aspirates.

–fluorescent antibody reaction on clinical specimens.

–culture in vitro.

B. Clinical Specimens

–extremely contagious; warn laboratory personnel that specimen contains suspected *Y. pestis*.
–for plague, culture sputum, node aspirate, blood, throat.
–for yersiniosis, culture stool, blood, serous cavities, mesenteric lymph nodes.

V. Control

A. Treatment

Y. pestis

–urgent to begin streptomycin intravenously because of rapid progression (pneumonic form is untreatable 12 to 15 hours after first symptom).
–untreated bubonic is 50% to 75% fatal; untreated pneumonic is 99% fatal.
–quarantine patients.

Yersiniosis

–perform antibiotic sensitivity on isolated organisms.
–monitor fluids and electrolytes.
–untreated can be 50% fatal.

B. Prevention

Y. pestis

–give prophylactic tetracycline to close contacts in previous 7 days.
–90% of world's plague is in southeast Asia; a very few cases of bubonic plague have been reported in southwestern U.S.
–vaccine is effective only against bubonic plague; it provides short-term protection; give to travelers to southeast Asia endemic areas.

VI. Virulence Attributes

Y. pestis

–multiplies intracellularly within monocytes; it also multiplies extracellularly.
–at least five different antigens are associated with virulence.

Yersiniosis

–enterotoxin similar to *E. coli*.

Francisella

I. General Comments

–tularemia is also termed deerfly fever or rabbit fever.
–zoonosis primarily in rabbits and rodents.
–human transmission by eating/handling infected animals or by bite of ticks, deerflies, blackflies, mosquitos, mites, lice; rabbits and ticks are main sources of infection.

II. Classification

–*F. tularensis*.

III. Clinical Manifestations

–characterized by macrophage infiltration, granulomas, necrosis of infected tissues; regional nodes become infected and suppurate; frequent spread to lungs, liver, and spleen.

–initial manifestations involve abrupt onset of fever, headache, and regional (painful) adenopathy; back pain, anorexia, chills, sweats, and prostration follow; 1% fatality rate.
–manifestations vary according to site of entry:

A. Ulceroglandular involves skin infection (most common form)
B. Ocularglandular involves eye infection with granulomatous conjunctivitis
C. Glandular or pneumonic involves lung infection from inhalation of infected dust
D. Typhoidal involves GI tract from ingestion of contaminated meat

IV. Laboratory Diagnosis

A. Identification

–gram-negative coccobacillus.
–very difficult to culture in vitro.
–agglutination reaction to detect rise in antibody titre in acute versus convalescent sera.
–skin test is available.

B. Clinical Specimens

–very contagious after in vitro growth, so do not routinely ask laboratory to culture.

V. Control

A. Treatment

–streptomycin; if early in infection, rapid cure is anticipated; if later in infection, it is more difficult to cure.
–relapses may occur due to intracellular residence of organisms.

B. Prevention

–thoroughly cook meat, especially rabbit meat.
–vaccine is available for high-risk groups such as sheep handlers, trappers, and laboratory workers.

VI. Virulence Attributes

–intracellular residence within fixed macrophages of reticuloendothelial system and within mononuclear phagocytes.

Brucella

I. General Comments

–zoonosis with most human infections in livestock farmers and meat processors.
–intracellular parasites of the reticuloendothelial system.

II. Classification

–three species cause human brucellosis:

B. suis
B. melitensis
B. abortus

III. Clinical Manifestations

–organisms localize and cause granulomas in spleen, liver, bone marrow, and lymph nodes.

–patient presents with intermittent fever and nondescript findings: profound muscle weakness, chills, sweats, anorexia, headache, backache, depression, nervousness.
–two types of infection:

A. Acute with relapses and fever
B. Chronic with protracted weakness, depression, arthralgias, myalgias lasting over 12 months

IV. Laboratory Diagnosis

A. Identification

–based primarily on prolonged clinical manifestations and serologic agglutination reaction.
–in vitro culture is not usually successful unless specimens are obtained very early in infection.
–gram-negative coccobacilli (stain after in vitro culture).

B. Clinical Specimens

–blood provides best chance for culture.
–tissue biopsies of spleen, bone marrow, liver, lymph nodes.

V. Control

A. Treatment

–tetracycline *and* streptomycin for 3 to 6 weeks (hard to eradicate because of intracellular residence).
–relapses in 30% of cases.

B. Prevention

–most infections in U.S. attributed to contaminated milk and cheese; routine pasteurization has greatly reduced incidence.
–effective vaccine only for animals.
–continued problem for farmers and meat processors.

VI. Virulence Attributes

–intracellular multiplication in macrophages of reticuloendothelial system; initial exposure leads to phagocytosis by PMNs, which carry organisms to lymph nodes, spleen, bone marrow, and liver with ensuing infection of these tissues.

Bacillus

I. General Comments

–zoonosis especially prevalent in sheep and cattle.
–farmers and animal processors are at risk; very low incidence in U.S.
–spores play an important role; they can survive in soil for 30 years.

II. Classification

–*B. anthracis* causes anthrax.
–*B. cereus* causes outbreaks of food poisoning.
–numerous other nonpathogenic bacillus species that occasionally cause problems in compromised hosts.

III. Clinical Manifestations

B. anthracis

–is essentially a toxemia in which a potent exotoxin enters the circulation and causes profound toxemia and death within 2 to 5 days.

–three types of manifestations depending on the route of entry of the spores:

A. Cutaneous (95% of anthrax) - spores enter cut or abrasion especially on hands, forearms, or head; small pustule develops into a large vesicle containing dark fluid (eschar) surrounded by a characteristic inflammatory ring at base; 10% fatal

B. Pulmonary (5% of anthrax - "wool sorter's disease") - abrupt onset of high fever, malaise, cough, myalgias; marked hemorrhagic necrosis of lymph nodes; respiratory distress and cyanosis; 85% fatal

C. GI (less than 1% of anthrax) - nausea, vomiting, bloody diarrhea; profound prostration and shock; 50% fatal

B. cereus

–two types of gastroenteritis: emetic involving severe nausea and vomiting (frequently confused with staphylococcal food poisoning) or diarrheal involving abdominal pain and profuse watery stools (frequently confused with clostridial food poisoning).

IV. Laboratory Diagnosis

A. Identification of Anthrax

–gram-positive large rod (spores are not present in clinical specimen).
–blood agar to isolate colonies.
–antibody tests on acute and convalescent sera.
–demonstrate toxicity by injecting mice; death in a few days with large numbers of bacilli in blood.

B. Clinical Specimens for Anthrax

–blood; if organisms are detected, prognosis is poor.
–sample of vesicle fluid.

V. Control

A. Treatment of Anthrax

–early detection is key to diagnosing cutaneous anthrax; pulmonary and GI anthrax are usually diagnosed postmortem.
–give penicillin intravenously to kill organisms and stop the elaboration of the exotoxin.
–if misdiagnosed as staphylococcal lesion and attempt to drain, organisms will be driven into bloodstream with disastrous consequences.

B. Prevention of Anthrax

–kill animals and bury deeply (spore problem); do not incinerate animals because spores are released into air.
–in endemic areas (Louisiana, Texas, California, South Dakota, Nebraska) gas sterilize commercial wool, hair, and hides.
–effective vaccine routinely used in animals where outbreaks have occurred.

VI. Virulence Attributes

B. anthracis

–capsular polypeptide of D-glutamic acid (unique for bacteria that usually contain polysaccharide capsules) inhibits phagocytosis.

–potent exotoxin causes CNS distress with respiratory failure and anoxia; composed of protective antigen, lethal factor, and edema factor.

–spores are phagocytosed by macrophages that carry them to nodes and induce infection.

B. cereus

–enterotoxin causes gastroenteritis.

Review Test

ZOONOSES

DIRECTIONS: For each of the questions or incomplete statements below, *one* or *more* of the answers or completions given is correct. Choose answer:

> A. if only **1, 2,** and **3** are correct
> B. if only **1** and **3** are correct
> C. if **2** and **4** are correct
> D. if only **4** is correct
> E. if all are correct

2.1. *Listeria monocytogenes* infections:

1. abscesses and granulomas within various tissues produce protean manifestations.
2. laboratory isolation involves cold enrichment techniques.
3. neonatal infections have very poor prognosis.
4. bacteria parasitize PMNs and monocytes.

2.2. *Yersinia pestis* infections:

1. wild rodents and their infected fleas are the primary reservoir.
2. sudden onset of fever with conjunctivitis along with regional bubo formation are key early manifestations.
3. pneumonic form is far more contagious than bubonic form.
4. bubonic form may progress to pneumonic form.

2.3. *Francisella tularensis* infections:

1. if suspected, ask hospital laboratory to culture.
2. effective vaccine is available; it is used only for high-risk groups.
3. intracellular parasite of PMNs.
4. relapses may occur after treatment.

2.4. Yersiniosis infections:

1. severe GI disease with characteristic mesenteric lymph node involvement.

2. termed "black death."
3. septicemia occurs primarily in debilitated patients.
4. caused by *Y. pestis* and *Y. enterocolitica*.

2.5. Brucella infections:

1. *B. suis* is the human pathogen; *B. melitensis* and *B. abortus* cause infections in animals only.
2. profound muscle weakness along with other relatively nondescript manifestations.
3. zoonotic disease spread to humans primarily by rat fleas.
4. patients with untreated infections may exhibit manifestations for as long as 1 year.

2.6. Anthrax infections:

1. early cutaneous form may mimic staphylococcal furuncle or boil.
2. pulmonary and GI anthrax are rapidly progressive with high fatality rates and are usually diagnosed postmortem.
3. spores play a key role in transmitting infections to humans.
4. capsule, like most bacteria, is long-chain polysaccharide.

Answers and Explanations

ZOONOSES

2.1. A. Answer 4 is not correct. These bacteria parasitize only monocytes and not PMNs.

2.2. E. All answers are correct.

2.3. C. Do not ask the hospital laboratory to culture; this organism is highly contagious after in vitro growth and only specially equipped laboratories should attempt culture. These bacteria parasitize fixed macrophages and mononuclear phagocytes.

2.4. B. "Black death" applies only to plague. *Y. pestis* is not considered a yersiniosis; yersiniosis includes *Y. enterocolitica* and *Y. pseudotuberculosis*.

2.5. C. All three species of brucella are human pathogens. Brucella is spread to humans primarily by handling infected meat.

2.6. A. Capsule of *B. anthracis* is unique for bacteria in that it is a polypeptide of glutamic acid.

Treponema

I. General Comments

–treponema are corkscrew-shaped, motile organisms.

–chronic, painless infections that may last 30 to 40 years if untreated.

–well-defined clinical stages; as host defenses are stimulated, treponemal numbers decline and patient becomes asymptomatic; this is followed by treponemal multiplication with subsequent reemergence of symptoms.

–syphilis is sexually transmitted; yaws and pinta are not.

–unusual bacterial morphology of outer envelope, three axial filaments, cytoplasmic membrane-cell wall complex, protoplasmic cylinder.

II. Classification

–*T. pallidum* subspecies *pallidum* causes syphilis (epidemic worldwide).

–*T. pallidum* subspecies *pertenue* causes yaws (hot tropical climates, not in U.S.).

–*T. pallidum* subspecies *carateum* causes pinta (Central and South America, not in U.S.).

–all three subspecies are morphologically and antigenically identical; differentiation is based solely on clinical manifestations.

III. Clinical Manifestations

–vascular involvement with endarteritis and periarteritis leading to inhibited blood supply and necrosis.

–lymphocyte and plasma cell infiltration at sites of infection.

–pathogenesis varies considerably: syphilis involves every tissue of the body; yaws involves bones and soft tissues; pinta involves skin only.

A. Syphilis

–primary stage: localized infection with erythema, induration with firm base (hard chancre - *Hemophilus ducreyi* is soft chancre), ulceration.

–secondary stage: disseminated infection with lesions in almost every tissue; mucocutaneous rash; may recur if untreated.

–tertiary stage: aortitis and CNS problems may kill patient.

–congenital infection during pregnancy: abortion, stillbirth, or birth defects following in utero infection.

B. Yaws

–primary stage: localized infection involving red papule that ulcerates.

–secondary stage: generalized spreading lesions in successive crops over months or years.

C. Pinta

–epidermal and dermal infection with eventual depigmentation of skin.

IV. Laboratory Diagnosis

A. Identification

–syphilis and yaws

1. clinical manifestations
2. darkfield microscopy of lesion exudate to demonstrate corkscrew-shaped spirochete (organisms too thin to gram stain)

3. serologic reactions - watch for biologic false positives that may confuse the diagnosis (positive serology in absence of treponemal disease)

–pinta differentiated by geographic location and clinical manifestations.

B. Clinical Specimens

–lesion exudate from pustule or ulcer for darkfield microscopy.
–cannot grow in vitro.

V. Control

A. Treatment

–long-acting penicillin.
–Jarisch-Herxheimer reaction immediately following antibiotic therapy for secondary syphilis involves intensification of manifestations that lasts for 12 hours; this is a good indicator that penicillin is effective.
–serology becomes negative 6 months after primary syphilis and 12 months after secondary syphilis; beyond secondary stage, patient may remain seropositive for many years.

B. Prevention

–use condom to prevent transmission.
–treat all sexual contacts prophylactically with penicillin.
–during pregnancy, do syphilis serology in both first and third trimester.

VI. Virulence Attributes

–immunosuppressive treponemal components are responsible for the chronic nature and for subsequent emergence of different stages.

Borrelia

I. General Comments

–relapsing fever is extremely rare (few cases/year).
–Lyme disease is rapidly increasing; causative agent was identified in 1984, and since then, alarming yearly increases.
–zoonosis in which humans are accidentally infected.
–unusual bacterial morphology of outer envelope, 15 to 22 axial filaments, cell membrane-cell wall complex, and protoplasmic cylinder.

II. Classification

–*B. burgdorferi* causes Lyme disease.
–*B. recurrentis* causes epidemic relapsing fever and is transmitted by lice.
–*Borrelia* . . . (species name depends on tick species) causes endemic relapsing fever and is transmitted by ticks.

III. Clinical Manifestations

A. Lyme Disease

–bloodstream infection seeds other tissues, especially nerves, heart, and joints.
–manifestations are protean, and this disease, like syphilis, has been termed "the great imitator."
–three distinct stages with considerable overlap; not all untreated patients exhibit all three stages.

–Stage I: erythema chronicum migrans is hallmark (intense circular rash that spreads out from the tick bite site); may last for several weeks and enlarge to several centimeters in diameter; malaise, fatigue, headache, fever, chills, stiff neck, aches, and pains for several weeks.

–Stage II: neural and heart problems; meningitis, cranial neuropathy, radiculoneuropathy, some cardiac dysfunction; follows stage I weeks to months later.

–Stage III: joint problems especially in large joints, producing oligoarthritis; bouts of arthritis for 3 to 7 years; neural dysfunction leads to dementia and paralysis; follows stage I months to years later.

B. Relapsing Fever

–bloodstream infection producing *sudden onset* of high fever, chills, headache, drenching sweats.

–after a few days, patient becomes afebrile for several days to several weeks.

–similar manifestations reappear and patient may exhibit four to ten consecutive relapses.

IV. Laboratory Diagnosis

A. Identification

–culture of Borrelia is rarely successful.

–key to diagnosis is clinical manifestations along with recent history of tick exposure.

–for Lyme disease, erythema chronicum migrans is highly characteristic (25% to 50% of patients may not experience this unique rash).

–for relapsing fever, relapses are highly characteristic.

–antibody detection is used to confirm Lyme disease or relapsing fever.

B. Clinical Specimens

–for Lyme disease, test CSF and blood; titers should be higher in CSF.

–for relapsing fever, blood specimens for antibody titers.

V. Control

A. Treatment

–penicillin or tetracycline for Lyme disease; progressively harder to eradicate in stages II and III; for stage III azithromycin has been effective where other antibiotics have failed.

–tetracycline for relapsing fever.

B. Prevention

–endemic areas for Lyme disease are Minnesota, Wisconsin, and northeastern seaboard.

–highest incidence of Lyme disease correlates with tick season (summer).

–if tick is attached and imbedded, carefully remove to avoid leaving reminants of mouth parts in skin.

VI. Virulence Attributes

–relapsing fever spirochete exhibits characteristic surface antigenic variation; antibodies clear organisms and variant organisms arise causing the next relapse.

Leptospira

I. General Comments

–zoonotic disease in which humans are infected by contact with urine of wild rodents and domestic animals.
–protean manifestations.
–cause Weil's disease, Fort Bragg fever, canicola fever, swineherd's disease.
–unusual bacterial morphology of outer envelope, two axial filaments, cytoplasmic membrane-cell wall complex, protoplasmic cylinder.

II. Classification

–*Leptospira interrogans* is the pathogen; 180 different serotypes or serovars.
–*Leptospira biflexa* is the nonpathogen.

III. Clinical Manifestations

–organisms enter through mucous membranes or break in skin; gain access to bloodstream and localize primarily in kidney, liver, and CNS.
–biphasic disease: first week in blood is acute stage; this is followed by short asymptomatic period then chronic stage.
–acute icteric disease: *abrupt onset* of fever, chills, headache, myalgias, GI upset, conjunctival suffusion, meningismus ("aseptic" meningitis).
–organisms in liver cause jaundice; organisms in kidney cause nephritis.
–fatalities result from kidney failure.

IV. Laboratory Diagnosis

–suspect in any case of acute fever of unknown origin.

A. Identification

–spirochete with hooked ends (treponemes do not have hooked ends).
–difficult to grow in culture; use special media and incubate 30°C; if growth occurs, it may require up to 4 weeks of incubation.
–agglutination reactions after growth in vitro.

B. Clinical Specimens

–acute phase, culture blood and CSF.
–chronic phase, culture urine.

V. Control

A. Treatment

–antibiotic treatment required within first few days of manifestations; if antibiotic started after 4 days or more of manifestations, little effect; problems are the generalized nature of manifestations along with lack of suspicion of leptospirosis.
–penicillin is the drug of choice.
–if jaundice, watch kidney function.
–supportive care: bed rest, relieve headache and muscle aches, watch fluids and electrolytes.

B. Prevention
–vaccinate pets and livestock (not humans).
–avoid contact with animal urine.

VI. Virulence Attributes
–poorly defined.

Review Test
SPIROCHETES

DIRECTIONS: For each numbered item, select one lettered heading that is most closely associated with it.

A. Treponema
B. Borrelia
C. Leptospira
D. all are correct
E. none are correct

2.1. gram-positive cocci.

2.2. relapsing fever transmitted by lice or ticks.

2.3. may present as acute icteric disease with protean manifestations.

2.4. zoonotic disease transmitted to humans primarily by contact with animal urine.

DIRECTIONS: For each of the questions or incomplete statements below, *one* or *more* of the answers or completions given is correct. Choose answer:

        A. if only **1, 2,** and **3** are correct
        B. if only **1** and **3** are correct
        C. if **2** and **4** are correct
        D. if only **4** is correct
        E. if all are correct

2.5. Clinical manifestations of Lyme disease:

1. termed the "great imitator" because of so many different manifestations.
2. the three well-defined stages do not overlap.
3. the causative agent is *Borrelia burgdorferi*.
4. stage III occurs very rapidly; manifestations may appear within 4 weeks of the initial infection.

2.6. Treatment and prevention of Lyme disease:

1. all three stages, like syphilis, are relatively easy to cure with the appropriate antibiotic therapy.
2. the endemic areas in the U.S. are the West Coast states.
3. spread to humans occurs through flies or lice.
4. highest incidence of the initial stage I disease occurs in the summer.

2.7. Congenital syphilis:

1. perform syphilis serologic test in first and third trimester of pregnancy.

2. diagnose via clinical manifestations, darkfield microscopy of lesion material, and serologic reactions.
3. congenital infection may produce abortion, stillbirth, or birth defects.
4. organisms do not infect in utero; newborns are infected as they pass through the infected birth canal.

2.8. Clinical manifestations of syphilis:

1. vascular involvement is characteristic.
2. termed soft chancre (*Hemophilus ducreyi* causes hard chancre).
3. secondary stage infects most tissues of the body.
4. secondary stage is rarely fatal.

2.9. Syphilis, yaws, and pinta infections:

1. chronicity is characteristic.
2. differentiation of these three diseases is based solely on clinical manifestations.
3. all three diseases are caused by treponemes.
4. all three diseases are sexually transmitted.

Answers and Explanations

SPIROCHETES

2.1. E. Spirochetes have a unique bacterial morphology; they are long, spiral-shaped organisms that have an outer envelope, axial filaments, cytoplasmic membrane-cell wall complex, and a protoplasmic cylinder.

2.2. B. Borrelia.

2.3. C. Leptospira.

2.4. C. Leptospira.

2.5. B. The three stages of Lyme disease have frequent overlap. Stage III manifestations are relatively slow to develop in this chronic disease; stage III may not emerge for several months or years after the initial infection.

2.6. D. First three answers are not correct. Stage III can be extremely difficult to eradicate. Lyme disease is endemic in Minnesota, Wisconsin, and the eastern seaboard states. Spread to humans is primarily through tick bites.

2.7. A. *T. pallidum* readily penetrates the maternal-fetal barrier to infect in utero.

2.8. B. *T. pallidum* causes hard chancre, and *H. ducreyi* causes soft chancre. Secondary syphilis is never fatal; tertiary stage kills.

2.9. A. Only syphilis is sexually transmitted; yaws and pinta are spread by casual contact.

Mycobacteria

I. General Characteristics

–have a complex peptidoglycan-arabinogalactan cell wall that is about 60% lipid.

–are resistant to acid and alkali, which allows NaOH treatment of sputa to reduce normal contaminating bacteria.

–are mostly very slow growers. *M. leprae* cannot be cultured in vitro.

–are resistant to drying and many disinfectants.

–stimulate a strong cell-mediated immune response in a healthy host.

–are acid-fast bacilli retaining dyes even when decolorized by acid alcohol because of long-chain fatty acids called the mycolic acids in the cell wall.

II. Classification

–are related to the Corynebacteria and Actinomycetes.

–include *Mycobacterium tuberculosis* and *M. bovis*, which cause tuberculosis.

–include *Mycobacterium leprae*, which causes leprosy.

–include a number of nontuberculous Mycobacteria, which cause disease only when introduced in large numbers or into a compromised host.

–are speciated based on growth rate and chemical characteristics such as levels of catalase production, niacin production, nitrate reduction, pigment production in the light and dark, arylsulfatase, Tween 80 hydrolysis, tellurite reduction, tolerance to salt, and growth on MacConkey agar.

–can be typed using specific mycophages.

III. Clinical Disease

A. Tuberculosis

–is caused by *Mycobacterium tuberculosis* and *Mycobacterium bovis*; attributes of virulence include:

sulfatides (sulfur containing glycolipids), which potentiate the toxicity of cord factor and promote intracellular survival by inhibiting phagosome-lysosome fusion.

cord factor (a trehalose mycolate), which disrupts mitochondrial membranes, interfering with respiration and oxidative phosphorylation; inhibits PMN migration; causes the organism to grow in culture in a serpentine or cord fashion; and is associated with granuloma formation.

–exposure occurs most commonly through inhalation of organisms in droplet nuclei from another individual or through ingestion of contaminated food.

–is more common in lower socioeconomic classes, recent Asian immigrants (where drug resistance is often a problem) and in AIDS patients.

–infection begins in the lungs and, in the immunologically naive, with a mild inflammatory reaction. The phagocytosed organisms replicate rapidly leading to the formation of a small Ghon lesion in the lung and the regional lymph nodes (which together are called the Ghon complex). Clinical symptoms are usually not significant at this stage.

–infection then proceeds with the development of the cell-mediated immune responses to fibrose, calcify, and heal (although the organism remains viable without treatment) in a healthy individual or, in an individual who is less immunologically competent, to spread. The spread is either by coalescence or by the hematogenous route (which results in miliary TB).

–infection may be reactivated in an individual who has "healed" primary tuberculosis lesions but was not treated. This often does not happen until after the fifth decade of life or under immunocompromising conditions. This is known as secondary or postprimary tuberculosis and occurs only in previously sensitized individuals. Necrosis is prominent.

–is diagnosed by:

1. typical clinical symptoms of cough and abnormal chest x-ray, and
2. demonstration of acid-fast bacteria in sputum, induced sputum, or gastric washings, *and*
3. culture of *M. tuberculosis* or *M. bovis* from specimens.
4. skin testing with purified protein derivative of *M. tuberculosis* (PPD):
   a. known recent conversion from − to + indicates infection.
   b. individuals with positive test of unknown duration have been infected but may not have active disease.
   c. individuals with known tuberculous disease and a negative PPD are anergic to the antigen, a bad prognostic sign.

–is treated prophylactically (known exposure with conversion of skin test to positive) with isoniazid.

–is treated in uncomplicated pulmonary tuberculosis in a previously untreated, cooperative patient with isoniazid and rifampin with ethamutol added if patient is from an area with drug resistance, such as Asia.

B. Leprosy

–is an intracellular infection of human beings with the invasion of nerve cells.

–is caused by *Mycobacterium leprae,* which is an obligate intracellular parasite of human beings.

–has two extreme forms on the ends of the spectrum of the disease.

1. Tuberculoid leprosy, in which cell-mediated immunity is functioning (lepromin test positive), and there are 1 to 3 bronze-colored cutaneous lesions with marked nerve anesthesia and enlargement; few organisms in skin.
2. Lepromatous leprosy, in which cell-mediated immunity fails, lepromin test is negative, and there are numerous skin lesions, which are macular to nodular; nerve anesthesia is less notable; tissues are loaded with *M. leprae,* particularly the mononuclear or epitheloid cells (known then as lepra cells); is the most contagious of the extremes.

–is diagnosed by cytology and lepromin (skin) test (not culture).

–is treated with dapson and rifampin.

–is prevented by isolation of actively infectious cases, by prophylactic treatment of close contacts, and by Bacille Calmette-Guérin (BCG) vaccination of children in villages where leprosy is a problem.

C. Nontuberculous Mycobacterial Diseases

–include pulmonary disease (in persons with chronic lung disease, cancer, or AIDS) most commonly caused by *M. kansasii* or *M. avium-intracellulare* but also many others.

–include mycobacterial lymphadenitis (most often in children) commonly caused by *M. scrofulaceum.*
–include soft tissue infection in tropical fish enthusiasts (*M. marinum, M. ulcerans*) or in surgical wounds *(M. fortuitum-chelonei).*
–include disseminated disease in immunocompromised individuals.
–are caused by 20 or so species of mycobacteria (excluding *M. tuberculosis, M. bovis* and *M. leprae*), which are speciated as discussed in the general characteristics.
–were formerly called atypical mycobacterial diseases.
–are most properly called "mycobacterial disease caused by *M.* _____" (and list species).
–are not considered contagious but are contracted from environmental sources.
–are diagnosed by repeated isolation of organisms by culture.
–are in many cases difficult to treat, depending on the species and the underlying state of health of the patient.
–are generally treated with the same antituberculous drugs.

Nocardia

I. General Characteristics

–is gram positive and partially acid-fast.
–may branch and then convert to rods.
–grows relatively slowly.

II. Clinical Diseases

A. Pulmonary Infection with Nocardia

–may metastasize to brain.
–is acute in children and compromised adults, or chronic in other adults.
–is often diagnosed at autopsy.
–is diagnosed by gastric washings, lung biopsy, and brain biopsy with culture and cytology.
–is treated by surgical drainage of abcesses and removal of necrotic tissues and sulfonamides.

B. Mycetoma as per *Actinomyces*, except treatment is with sulfonamides.

Actinomyces

I. General Characteristics

–are non-acid fast.
–are anaerobic.
–may branch in tissue and then convert to rod forms.
–cause chronic granulomatous infections mainly of the soft tissues (although it will invade bone) with swelling and a tendency to form sinus tracts to the surface that contain hard microcolonies called granules.

II. Classification

–are related to the mycobacteria and belong to the Actinomycetes.

III. Attributes of Virulence
–grows contiguously without respect for anatomic barriers.

IV. Clinical Disease

A. Cervical Facial or Lumpy Jaw
–begins most commonly following dental work.

B. Thoracic
–may involve lungs and ribs.

C. Abdominal
–starts in ileocecal region and frequently produces sinus tracts to skin surface.

D. Mycetoma
–is an infection of limb with swelling, sinus tract formation, and granules.

V. Diagnosis
–is by Gram's stain of granules and anaerobic culture.

VI. Treatment
–is commonly by penicillin treatment and surgical drainage of necrotic tissues.

Review Test

MYCOBACTERIA, NOCARDIA, ACTINOMYCES

DIRECTIONS: For each of the questions or incomplete statements below, one or more of the answers or completions given is correct. Choose answer:

    A. if only **1, 2,** and **3** are correct
    B. if only **1** and **3** are correct
    C. if **2** and **4** are correct
    D. if only **4** is correct
    E. if all are correct

2.1. Comparing tuberculosis with nontuberculous pulmonary disease.

1. TB is as contagious as nontuberculous pulmonary disease.
2. Both are caused by species of the genus *Mycobacterium.*
3. TB is caused only by *Mycobacterium tuberculosis*; nontuberculous pulmonary disease by numerous other species.
4. Both are being seen in increasing frequency in AIDS patients.

2.2. True statements about various actinomycetes:

1. *Nocardia* is partially acid-fast and filamentous.

2. *Actinomyces* are partially acid-fast and filamentous.
3. *Mycobacterium leprae* must be cultured on special medium such as Lowenstein-Jensen medium.
4. *Mycobacterium* is generally rod shaped and acid-fast.

2.3. Mycolic acids:

1. Are found in the mycobacterium cell wall.
2. Are part of the cord factor.
3. Are responsible for the acid-fastness of mycobacteria.
4. Are long-chain fatty acids.

DIRECTIONS: Each of the questions or incomplete statements below is followed by suggested answers or completions. Select the *one best* in each case.

2.4. Which of the following grows contiguously in tissues without respect to anatomic barriers and thus often invades bone?

A. *Actinomyces israelii.*
B. *Nocardia asteroides.*
C. *Mycobacterium leprae.*
D. *Mycobacterium kansasii.*
E. *Mycobacterium tuberculosis.*

2.5. The Ghon complex is found in:

A. Actinomycotic mycetoma.
B. Tuberculoid leprosy.
C. Primary tuberculosis.
D. Cervical facial actinomycosis.
E. Soft tissue infections caused by *M. marinum.*

100

Answers and Explanations

MYCOBACTERIA, NOCARDIA, ACTINOMYCES

2.1. C. Nontuberculous mycobacterial pulmonary disease is not considered contagious. TB is caused by *M. tuberculosis* or *M. bovis*; nontuberculous disease by *M. avium-intracellulare* and others. Both forms are seen with increasing frequency in AIDS patients.

2.2. D.

	acid-fast	*shape*
Mycobacterium	+	rod
Actinomyces	−	filamentous → rod
Nocardia	±	filamentous → rod

2.3. E.   All are true statements.

2.4. A.   *Actinomyces* is most notorious although *Nocardia* may also cross anatomic barriers.

2.5. C.

3

Virology

Nature of Animal Viruses

I. Virus Particles

–are called virions.

–are composed of either RNA or DNA, which is encased in a protein coat called a *capsid*.

–are either naked or enveloped depending on whether the capsid is surrounded by a lipoprotein *envelope*.

–replicate only in living cells and are, therefore, obligate intracellular parasites.

–cannot be observed with a light microscope.

A. The Viral Genome

–may be single-stranded or double-stranded, linear or circular, and segmented or nonsegmented.

–is used as one criterion for viral classification.

–is associated with viral-specific enzymes and/or other proteins within the virion.

B. The Viral Capsid

–is composed of structural units called *capsomers*, which are aggregates of *viral-specific polypeptides*.

–has a symmetry that is classified as either *helical, icosahedral (20-sided polygon)*, or *complex*.

–is used as a criterion for viral classification.

–serves four functions: (1) protection of the viral genome, (2) site of receptors necessary for naked viruses to initiate infection, (3) stimulus for antibody production, and (4) site of antigenic determinants important in some serologic tests.

C. The Viral Nucleocapsid

–refers to the capsid and enclosed viral genome.

–is identical to the virion in naked viruses.

D. The Viral Envelope

–surrounds the nucleocapsid of enveloped viruses.

–is composed of *viral-specific glycoproteins* and *host-cell derived lipids and lipoproteins*.

–contains molecules that are (1) necessary for enveloped viruses to initiate infection, (2) a stimulus for antibody production, and (3) antigens in serologic tests.

–is the basis of ether sensitivity of a virus.

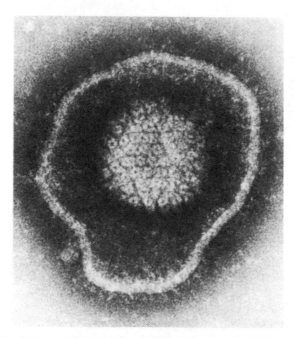

Figure 3.1. A single herpesvirus particle that has degraded sufficiently to allow visualization of the internal capsid. The complete particle now measures over 200 nm in diameter, and the capsid is 100 nm. The icosahedral form of the capsid is apparent, and there are 162 capsomeres on a P = 1 icosahedron. × 275000. (From Brown F, Wilson GS, eds. Topley and Wilson's Principles of Bacteriology, Virology and Immunity. 7th ed. Baltimore: Williams & Wilkins, 1984, vol 4, p 42.)

II. Viral Classification

–is based on chemical and physical properties of virions.

–has resulted in the separation of viruses into *major families*, which are further subdivided by physiochemical and serologic characteristics into *genera*.

III. DNA Viruses

–contain double-stranded DNA except for parvovirus.

–are naked viruses except for herpesviruses, poxviruses, and hepadnaviruses.

–have icosahedral capsids and replicate in the nucleus except for poxviruses.

IV. RNA Viruses

–contain single-stranded RNA except for reoviruses.

–are enveloped except for picornaviruses and reoviruses.

–have helical capsids except for picornaviruses, reoviruses and togaviruses.

–replicate in the cytoplasm except for orthomyxovirus and retroviruses, which have both a cytoplasmic and a nuclear phase.

Viral Replication and Genetics

I. Viral Replication

–occurs only in living cells.

–involves many host cell enzymes and functions.

–follows a sequential pattern that involves the following steps: (1) attachment; (2) penetration; (3) uncoating viral genome; (4) synthesis of early proteins involved in genome replication; (5) synthesis of late proteins, which are structural components of the virion; (6) assembly; and (7) release.

–may be incomplete in some cells (*abortive infection*) and yield *defective particles* (lack functional replication gene) in some cells.

–may lead to death of the host cell (*virulent viruses*) or may occur without apparent damage to the host cell (*moderate viruses*).

–is very similar for all viruses in a specific family.

A. Plaques and Pocks

–are focal areas of viral-induced cytopathology observed in tissue culture cells monolayers, and membranes of embryonated eggs respectively.

–are counted following application of serial dilutions of a virus suspension to susceptible cells in order to quantitate the number of infectious virions present.

B. One-Step Multiplication Curves

–can be obtained for animal viruses.

–are plots of time after infection versus number of viral plaques or pocks.

–show viruses have an *eclipse period* (time from start of infection to the first appearance of intracellular infectious virus) and a *latent period* (time from start of infection to the first appearance of extracellular virus).

–indicate that viruses take a much longer time (hours or days) to replication than phage.

C. Attachment

–involves the interaction of viral receptors and specific host cell receptor sites.

–plays an important role in viral pathogenesis since it determines *viral cell trophism*.

–may be inhibited by antibodies against viral receptors or cellular receptor sites.

D. Penetration

–occurs by pinocytosis or phagocytosis.

–can involve the fusion of the virus envelope with the host cell's plasma membrane.

E. Uncoating

–refers to the separation of the capsid from the viral genome.

–results in the loss of virion infectivity.

F. Budding

–is the process by which enveloped viruses obtain their envelope.

–is preceded by the insertion of virus-specific glycoproteins into a host cell's membranes.

–occurs most frequently at the plasma membrane, but also occurs at other membranes.

–confers infectivity to enveloped viruses.

II. DNA Viruses

A. Transcription

–occurs by host cell DNA-dependent RNA polymerases, except for virion-associated RNA polymerase of poxviruses.

–results in transcripts that must have a poly A tail and methylated cap added before translation.

–occurs in specific temporal pattern, i.e., immediate early, delayed early, and late m-RNAs.

—may be followed by *posttranscriptional processing* of primary m-RNA transcripts (late adenovirus transcripts).

—occurs in the nucleus except for poxviruses.

B. Translation

—occurs on cytoplasm polysomes.

—is followed by transport of newly synthesized proteins to the nucleus except for poxviruses.

C. Genome Replication

—is *semiconservative.*

—is performed by a DNA dependent DNA polymerase, which may be supplied by the host cell (adenoviruses) or virus-specified (herpesviruses).

—occurs following the synthesis of the early proteins.

D. Assembly

—occurs in the nucleus, except for poxviruses.

—is frequently an inefficient process that leads to accumulations of viral proteins that may participate in the formation of *inclusion bodies.*

III. RNA Viruses

A. The Viral Genome

—may be single-stranded or double-stranded and segmented or nonsegmented.

—is said to have *messenger (positive-sense) polarity* if it is single-stranded and can act as m-RNA (picornaviruses and retroviruses).

—is said to have *antimessenger (negative-sense) polarity* if it is single-stranded and is complementary to m-RNA (orthomyxoviruses and paramyxoviruses).

—is *ambisense* if it is a single-stranded molecule that has segments of messenger polarity and segments of antimessenger polarity.

B. Transcription

—involves an RNA dependent RNA polymerase for all viruses except retroviruses, which use a host cell DNA dependent RNA polymerase (see Retroviruses).

—involves a virion-associated enzyme (transcriptase) with negative-sense viruses.

C. Translation

—occurs on cytoplasmic polysomes.

—may result in the synthesis of a large polyprotein that is subsequently cleaved (*posttranslational processing*) into individual viral polypeptides (picornaviruses and retroviruses).

D. Genome Replication

—occurs in the cytoplasm except for orthomyxoviruses and retroviruses.

—is performed by a viral-specific *replicase enzyme* except for retroviruses.

—involves a *replicative intermediate* RNA structure for all single-stranded RNA genomes.

—is asymmetric and conservative for the double-stranded RNA genomes.

IV. Genetics

A. Phenotypic Mixing

—occurs when pairs of related viruses with similar, but distinct, surface antigens infect the same cell.

–results when surface antigens from two related viruses enclose the genome of one of the viruses.

B. Phenotypic Masking (Transcapsidation)

–occurs when pairs or related viruses infect the same cell.
–results when the genome of one virus is surrounded by the capsid and/or capsid and envelope of the other virus.

C. Complementation

–can occur when two mutants of the same virus or, less frequently, two mutants of different large DNA viruses infect the same cell.
–results when one mutant virus supplies an enzyme or factor that the other mutant lacks.

D. Genetic Reassortment

–can occur when two strains of mutants of a segmented RNA virus infect a cell.
–results in a stable change in the viral genome.

E. Viral Vectors

–can be constructed with recombinant DNA technology and allow gene transfer into cells.
–are usually defective viruses that cannot replicate but can infect cells.
–have been used for the production of some vaccines, e.g., hepatitis B vaccine.

Viral Pathogenesis

I. Viral Pathogenesis

is the process of disease production following infection.
–may lead to *clinical* or *subclinical* (asymptomatic) disease.
–is the result of several different viral and host factors.

A. Viral Entry into a Host

–occurs through the mucosa of the respiratory tract most frequently.
–may occur through the mucosa of the gastrointestinal or the genitourinary tract.
–can be accomplished by direct virus injection into the bloodstream via a needle or insect bite.

B. Asymptomatic Viral Disease

–may also be called a *subclinical infection* because no clinical symptoms are evident.
–is the result of the majority of viral infections.
–can stimulate humoral and cellular immunity.

C. Clinical Viral Disease

–is frequently dependent on the size of the viral inoculum.
–does not always follow infection and therefore is not an accurate index of viral infection.
–is the result of viral-induced cellular injury or cellular death, which leads to the physiologic changes in infected tissues.
–is much less frequent than inapparent infections.
–is associated with a particular *target organ* for a specific virus.

II. Viral Aspects of Pathogenesis

A. Viral Surface Receptors

–interact with cellular receptor sites to initiate infection.

–may react with specific antibodies (neutralizing antibodies) and become incapable of interaction with cellular receptor sites.

–can be inactivated by pH, enzymes, and other host biochemicals.

B. Viral Virulence

–refers to the ability of a particular viral strain to cause disease.

–is genetically determined.

–is a composite of all the factors that allow a virus to overcome host defense mechanisms and damage its target organ.

–is decreased with *attenuated strains* of virus.

III. Cellular Aspects of Pathogenesis

A. Cellular Receptor Sites

–interact with virion receptors to initiate infection.

–help determine *cell trophism* of viruses.

–may be determined by the stage of differentiation of a cell.

B. Cell Trophism

–refers to the propensity of a virus to infect and replicate in a cell.

–is largely determined by (1) the interaction of virus receptors and cellular receptor sites and (2) the ability of the cell to provide other components, i.e., substrates, enzymes essential for viral replication.

C. Target Organ

–refers to that organ responsible for the major clinical signs of a viral infection.

–is largely determined by viral virulence and cell trophism.

D. Cellular Responses to Viral Infection

–result in clinical disease.

–may include (1) *cytopathic effects,* (2) *cytolysis,* (3) *inclusion body* formation, (4) *chromosomal aberrations,* (5) *transformation,* and (5) *interferon synthesis* or may be inapparent.

E. Cytopathogenic Effects

–include (1) inhibition of host-cell macromolecular biosynthesis, (2) alterations of the plasma membrane and lysosomes, and (3) development of *inclusion bodies* (focal accumulation of virion or viral gene products).

–may occur without the production of infectious virus progeny.

–may aid in the identification of certain viruses, i.e., polykaryocyte formation by measles virus.

IV. Types of Infections

A. Inapparent Infections

–occur when an insufficient number of cells are infected to cause clinical symptoms.

–are synonymous with *subclinical disease.*

–can result in sufficient antibody stimulation to cause immunity from further infections.

–occur frequently when the virus inoculum is small.

—occur when the virus does not reach its target organ.

B. Acute Infections

—occur when clinical manifestations of disease are observed for a short time (days to weeks) after a short incubation period.
—have recoveries associated with the elimination of the virus from the body.
—are classified as *localized* or *disseminated,* depending on whether the virus must travel from its site of implantation to its target organ.
—may lead to *persistent* or *latent* infections.

C. Persistent Infections

—are associated with the continuing presence of infectious virus in the body for extended, perhaps life-long, periods.
—may or may not have clinical symptoms.
—may involve infected individuals known as *carriers.*
—provide constant viral antigenic stimulation that leads to high antibody titers for some antigens.

D. Latent Infections

—occur when the infecting virus persists in the body in a noninfectious form that can periodically reactivate to infectious virus that produces clinical disease.
—are synonymous with *recurrent disease.*
—produce antibody stimulus only during the initial (*primary*) infection and during *recurrent* episodes.
—can have subclinical reactivations.
—are difficult to detect in cells because viral antigen production is not detected and cytopathology is not observed during "silent" periods.

E. Slow Infections

—have a prolonged incubation period lasting months or years.
—do not have clinical symptoms during incubation.
—do produce some infectious viruses during incubation.
—are most often associated with chronic, progressive, and fatal viral diseases of the central nervous system, like kuru and Creutzfeldt-Jakob disease.

V. Patterns of Disease

A. Localized Disease

—occurs when viral multiplication and cell damage remain localized to the site of viral entry into the body.
—has a short incubation time.
—may have systemic clinical features, e.g., fever.
—is not associated with a pronounced viremia (virions in the blood).
—occur in the respiratory tract (influenza, common cold), alimentary tract (picornaviruses and rotaviruses), genitourinary tract (genital warts), and the eye (adenovirus).
—can be spread over the surface of the body to other areas where it also causes another localized infection (picornavirus conjunctivitis).
—induces a much weaker immune response than disseminated infections.

B. Disseminated Infections

—involve the spread of virus from its site of entry to a target organ.

–involve a *primary and perhaps a secondary viremia.*
–involve moderate (weeks) incubation times.
–have their main clinical symptoms associated with infection of one target organ, although infection of other organs is involved.
–generate a substantial immune response that frequently confers life-long immunity to the host.
–allow a longer period of time for the host's immune system to eliminate the virus infection.

1. Viral Dissemination

–is a major feature of disseminated infections.
–may involve virus travel in (1) *other cells* (red blood cells and mononuclear peripheral white blood cells), (2) the *plasma,* (3) *extracellular spaces,* and (4) *nerve fibers.*
–is prevented by viral-specific cytotoxic cells and neutralizing antibodies.

C. Congenital Infections

–refer to viral infections of a fetus.
–are the result of a maternal viremia.
–lead to *maldeveloped organs.*
–are very serious due to (1) the immaturity of the fetal immune system, (2) the placental barrier to maternal immunity, and (3) the undifferentiated state and rapid multiplication of fetal cells.

Host Defenses to Viruses

I. Host Defense Mechanisms

–are responsible for the self-limiting nature of most viral infections.
–have immune and nonimmune aspects.
–are operating during all stages of a virus infection.
–may contribute to the clinical pattern of disease (immunopathology).

II. Nonimmune Defenses

A. Innate Immunity

–includes the *anatomic* (dead cells of epidermis) and *chemical* (mucous layers) *barriers* that limit content of virus with susceptible cells.
–includes the complex parameters associated with the *age and physiologic state of the host.*

B. Cellular Resistance

–involves *nonpermissive cells,* which lack factors necessary for virus replication.
–involves *lack of receptor sites* for virus on cells.

C. Inflammation

–limits the spread of virus from an infection site.
–results in unfavorable environmental conditions (antiviral substances, low pH, elevated temperatures) for viral replication.

D. Interferon

–is a host-specific, viral-induced glycoprotein that inhibits viral replication (see Immunotherapy chapter).
–is not viral-specific but fairly species-specific.

–is the *first viral-induced defense mechanism* at the primary site of infection.

III. Humoral Immunity

–involves the production of antibodies against viral-specific antigen by "B" lymphocytes.

–is the defense mechanism most important to cytolytic viral infections accompanied by viremia and viral infections of epithelium surfaces.

–involves both *virus-neutralizing* and *nonneutralizing antibodies.*

A. Neutralizing Antibodies

–inhibit the ability of a virus to replicate.

–can inhibit viral attachment, penetration, and/or uncoating.

–can, with the aid of complement, induce lesions in the viral envelope.

–are most protective if present at the time of infection or during viremia.

B. Nonneutralizing Antibodies

–enhance viral degradation.

–act as opsonins to enhance phagocytosis of virions.

IV. Cell-Mediated Immunity

–involves cytotoxic *"T" lymphocytes*, antibody-dependent cell-mediated cytotoxicity *(ADCC)*, natural killer *(NK) cells*, and activated macrophages.

–involves soluble factors from T lymphocytes (*lymphokines*) and macrophages (*monokines*) that regulate cellular immune responses.

–is the defense mechanism most important to noncytolytic infection in which the virus-infected cell's membrane is antigenetically altered by the virus.

V. Viral-Induced Immunopathology

–can contribute to the disease process.

–can result from various immunologic interactions, including immediate hypersensitivity, antibody-antigen complexes (hepatitis B virus), and tissue damage due to cytotoxic cells or antibody and complement.

–is frequently observed in *persistent viral infections.*

VI. Viral-Induced Immunosuppression

–results when infecting viruses alter the immune responsiveness of lymphocytes or decrease the numbers of lymphocytes.

–can occur during cytolytic or noncytolytic infection.

–is frequently observed as a *transient consequence of disseminated viral infections* that involve lymphocyte infection by the virus.

Immunotherapy, Antivirals, and Interferon

I. Immunotherapy

A. Virus Vaccines

–are examples of *active immunization.*

–are very effective in preventing infections caused by viruses that have few antigenic types.

–may use either *live virus, killed virus, virion subunits,* or *viral polypeptides.*

1. Live Virus Vaccines

–use *attenuated virus strains* that are relatively avirulent.

–have the advantages that they can be administered in a single dose by the natural route of infection and can induce a wide spectrum of antibodies and cytotoxic cells.

–have disadvantages that include limited shelf life, the possible reversion to virulence, and the possible production of persistent infection.

–include vaccines for *measles, mumps, rubella, polio (Sabin), yellow fever, and some adenovirus strains.*

2. Killed Virus Vaccines

–are prepared from whole virions by *heat or chemical inactivation* of infectivity.

–are injected into the body and stimulate antibodies only to the surface antigens of the virus.

–have the advantage of easily being combined into *polyvalent vaccines* (vaccines containing virions from several virus strains).

–have disadvantages that include no development of secretory IgA, boosters are needed, cell-mediated response is poor, and hypersensitivity reactions may occur.

–include vaccines for *poliovirus (Salk), rabies, and influenza.*

3. Virion Subunit Vaccines

–are purified proteins (viral receptors) obtained from virions.

–have the same advantages and limitations as killed vaccines.

–are available for *adenovirus.*

4. Viral Polypeptides

–are *polypeptide sequences of virion receptors* that have been synthesized or are from the purification of proteins made from cloned genes.

–have the same advantages and limitations as killed vaccines.

–are used in a vaccine for *hepatitis B virus.*

B. Passive Immunization

–refers to the injection of pooled human plasma or γ-globulin fractions from immune individuals into high-risk individuals.

–is valuable in the prevention of some viral disease but is of little value after onset of disease.

–is used to prevent *rubella, measles, mumps, hepatitis A and B virus, rabies, and varicella-zoster infections.*

II. Antivirals

A. Antiviral Drugs

–must selectively inhibit viral replication without affecting the viability or functioning of the host cell (selective toxicity).

–are of limited use today.

–include only *six licensed compounds:* acyclovir, amantadine, idoxuridine, ribavirin, trifluridine, and vidarabine.

B. Nucleoside Analogues

–constitute the majority of antivirals.

–are effective against herpesviruses.

–inhibit viral replication by inhibiting an enzyme involved in purine, pyrimidine, or DNA synthesis or are incorporated into DNA and inhibit its synthesis or function.

–are effective only on actively replicating virus (latent virus not affected).

1. Acyclovir

–is an analogue of guanosine.
–is phosphorylated by herpesvirus thymidine kinase.
–inhibits herpes viral DNA polymerase.
–helps control *herpetic eye infections and genital herpes reactivations.*
–suppresses reactivation of latent herpesvirus infections in immunosuppressed patients.

2. Vidarabine

–is a purine analogue, adenosine arabinoside (*Ara-A*).
–blocks viral DNA polymerase.
–inhibits both herpesvirus and poxvirus replication.
–is relatively nontoxic, but may cause nausea.
–is used to treat *herpetic encephalitis.*

3. Idoxuridine

–is a halogenated pyrimidine.
–inhibits both herpesvirus and cellular thymidine kinase.
–can be toxic to humans.
–was the first licensed antiviral for humans.

4. Trifluridine

–is a fluorinated thymidine molecule.
–is used for *herpetic keratitis* treatment.

5. Cytosine Arabinoside (ARA-C)

–is a pyrimidine analogue.
–inhibits viral and cellular DNA polymerase.
–is immunosuppressive and cytotoxic.

6. Ribavirin

–is a nucleoside related structurally to guanosine.
–is *active against DNA and RNA viruses* in vitro.
–affects the synthesis of functional viral m-RNA.

7. 3'-Azido-3'-Deoxythymidine (AZT)

–is a thymidine analogue.
–is converted to the triphosphate form by cellular enzymes and used by retrovirus reverse transcriptase.
–inhibits reverse transcriptase to a greater extent (100X) than the α cellular DNA polymerase.
–is presently in clinical trials for treatment of AIDS.

C. Other Types of Antivirals

–affect viral penetration, uncoating, or assembly.
–inhibit viral-specific enzymes.

1. Amantadine

–blocks penetration and uncoating of *influenza A virus.*
–may be chemically modified to *rimantadine,* which has less human side effects.

2. **Arildone (Win 51711 or 52084)**

    –stabilizes picornavirus virions and prevents uncoating.
    –inhibits picornavirus replication in tissue culture and animal models of human enterovirus disease.

3. **Methisazone (Marboran)**

    –blocks a late stage in the replication of *poxviruses*.
    –causes the formation of noninfectious virions.
    –was used to treat smallpox infections.

4. **Phosphonoacetic Acid (PAA) and Phosphonoformic Acid (PFA)**

    –inhibit herpesvirus DNA polymerase.
    –have been demonstrated to accumulate in bone.

III. Interferon

A. Interferons

–are host-coded *glycoproteins* that are produced in response to viruses, synthetic nucleotides, and foreign cells.
–are *host-specific* but not viral-specific.
–are divided into three groups or families: *IFN-α, IFN-β, and IFN-γ*.
–are produced in and secreted from virus-infected cells.
–bind to cell-surface receptors and *induce antiviral proteins*, including a protein kinase and "2,5 A" synthetase (synthesizes an oligoadenylic acid), which leads to the destruction of viral m-RNA.
–have toxic side effects including bone marrow suppression.

Diagnostic Virology

I. Laboratory Viral Diagnosis

–involves one of three basic approaches: (1) *virus isolation*; (2) *direct demonstration of virus, viral nucleic acid, or antigens* in clinical specimens; or (3) *serologic testing* of viral-specific antibodies.
–is frequently not necessary to diagnose a viral infection since the clinical symptoms are often distinctive.
–begins by identifying "most likely" viruses on the basis of clinical symptoms and patient's history.
–is often not possible during the first few days after infection.

II. Virus Isolation

–depends on virus replication in susceptible cells.
–may involve the use of tissue culture cells, embryonated eggs, or animals hosts.
–demands that clinical specimens are obtained and preserved correctly.
–is best accomplished during the onset and acute phase of disease.

A. Viral Replication

–may be detected in live infected tissue culture cells by observing a characteristic cytopathogenic effect (*CPE*), e.g., polykaryocyte formation or hemadsorption (adhesion of red blood cells to infected cells).
–may be observed in fixed infected tissue culture cells by (1) observing characteristic *inclusion bodies* or (2) immunohistochemical staining of *viral antigens*.

–is detected in embryonated egg by pock formation and in animals by the development of clinical symptoms.

III. Direct Examination of Clinical Specimens

–may be done on (1) sections of tissue biopsies, (2) tissue imprints or smears, (3) blood, (4) cerebrospinal fluid, (5) urine, (6) throat swabs, (7) feces, or (8) saliva.
–should be performed *only* on those specimens likely to contain the virus, i.e., throat swabs for respiratory tract infection.
–uses one of the following assays for virus detection: *viral-induced CPE, immunohistochemical staining, nucleic acid hybridization,* or *solid phase immunoassay.*

A. Immunohistochemical Staining

–uses fixed or fresh specimens and *chemically labeled (fluorescein)* or *enzymatically labeled (peroxidase) antibodies* to detect viral antigens.
–may employ "impression slides" made from specific tissues.
–may involve either a *direct or an indirect staining method.*

B. Nucleic Acid Hybridization

–involves the *detection of viral DNA or RNA sequences* in nucleic acid extracted from specimens.
–uses *"dot blot"* hybridization techniques that usually employ single-stranded complementary nucleic acid probes.
–is highly sensitive and specific.
–is a popular technique for the identification of *advenovirus* in nasopharyngeal washings and *cytomegalovirus* in urine.

C. Solid Phase Immunoassays

–are highly sensitive and specific assays used to *detect viral antigens.*
–employ specific viral antibodies and *radioimmunoassay (RIA)* or *enzyme-linked immunosorbent assay (ELISA) techniques.*
–are popular for the detection of *rotavirus* and *hepatitis A virus* in feces.

IV. Serologic Tests

–are used to determine the titer of specific antiviral antibodies.
–are helpful in diagnostic virology if *paired sera* are taken at the onset and recovery phases of the illness.
–must show at least a *four-fold rise in titer* between paired sera to indicate a current infection.
–may be diagnostic without the use of "paired sera" if significant levels of IgM antiviral antibodies are obtained.
–include *virus neutralization, complement fixation,* and *hemagglutination-inhibition tests,* and *solid-phase immunoassays.*

A. Virus Neutralization Tests

–are based on the principle that certain antiviral antibodies will neutralize the cytopathogenic effects of the virus.
–involve incubations of constant amounts of virus with decreasing amounts of serum, added to susceptible cells.
–are expensive to perform and must be standardized for each virus.

B. Hemagglutination-Inhibition Test

–is based on the principle that antihemagglutinin antibodies in serum will inhibit viral agglutination of erythrocytes.

–can be performed only on those viruses with hemagglutinins on their surface (influenza, measles).

–involve careful standardization of erythrocytes and viral hemagglutinin preparations.

C. Solid Phase Immunoassays

–are highly sensitive and specific assays used to *detect specific viral antibodies*.

–use viral antigens in RIA and ELISA protocols.

–are several hundred times more sensitive than other serologic tests.

Review Test

PRINCIPLES OF VIROLOGY

DIRECTIONS: Each of the questions or incomplete statements below is followed by suggested answers or completions. Select the *one best* in each case.

3.1. The eclipse period of a one-step viral multiplication curve is the time:

A. between the uncoating and assembly of the virus.

B. between the start of the infection and the first appearance of extracellular virus.

C. required for the synthesis of viral receptors.

D. between the start of infection and the first appearance of intracellular virus.

E. between the start of the infection and the uncoating.

3.2. Persistent virus infections:

A. are usually confined to the initial site of infection.

B. have no clinical symptoms during their incubation period.

C. elicit a poor antibody response.

D. may involve infected "carrier" individuals.

E. are associated with "localized" infections.

3.3. Which of the following host defense mechanisms is not viral-specific?

A. Neutralizing antibodies

B. Sensitized "T" cells

C. Complement-fixing antibodies

D. Interferon

E. None of the above

3.4. The most sensitive type of serologic test is:

A. Virus neutralization.

B. ELISA

C. Nucleic acid hybridization.

D. Hemadsorption.

E. Hemagglutination inhibition.

3.5. The initial host defense mechanism that occurs at the first site of primary virus infection is:

A. Inflammation.

B. IγM antibody production.

C. Sensitized "T" cell production.

D. Interferon production.

E. IγG antibody production.

DIRECTIONS: Each set of lettered headings below is followed by a list of numbered words or phrases. Choose answer.

A. if the item is associated with A only.

B. if the item is associated with B only.

C. if the item is associated with both A and B.

D. if the item is associated with neither A nor B.

A. Capsid
B. Envelope
C. Both
D. Neither

3.6. Contains viral-specific proteins and/or glycoproteins.

3.7. Can be dissociated by organic solvents.

3.8. Is used as a criteria for viral classification.

3.9. Contains a nucleic acid component.

117

3.10. Is partially derived from the host cell.

A. Genetic reassortment
B. Transcapsidation
C. Both
D. Neither

3.11. Results in a stable genetic change.
3.12. Occurs with both RNA and DNA viruses.
3.13. Occurs when related viruses infect the same cell.
3.14. Involves DNA ligase enzymes.
3.15. Occurs during the assembly stage of virus replication.

A. ELISA test
B. Immunohistochemical staining
C. Both
D. Neither

3.16. Is used to detect viral antigens.

3.17. Is used to classify viral inclusion bodies.
3.18. Is used to detect antiviral antibodies.
3.19. Depends on antibody reagents.
3.20. May be used in conjunction with "impression slides."

A. live enveloped virus vaccine
B. enveloped subunit virus vaccine
C. both
D. neither

3.21. Uses attenuated virus strains.
3.22. Stimulates antibodies to nucleocapsid antigens.
3.23. Stimulates humoral immunity.
3.24. Will not revert to a virulent ("wild-type") strain.
3.25. Stimulates cell-mediated immunity.

DIRECTIONS: For each of the questions or incomplete statements below, *one* or *more* of the answers or completions given is correct. Choose answer:

   A. if only **1, 2,** and **3** are correct
   B. if only **1** and **3** are correct
   C. if **2** and **4** are correct
   D. if only **4** is correct
   E. if all are correct

3.26. The infectious virus particle:
1. is an obligate intracellular parasite.
2. is called a viroid.
3. contains viral-specific enzymes.
4. may be inactivated by treatment with nucleases.

3.27. Viral transcription:
1. occurs in a specific temporal pattern for most DNA viruses.
2. involves a virion-transcriptase for negative-sense, enveloped RNA viruses.
3. involves host-cell RNA polymerases during herpesvirus replication.
4. may be inhibited by amantadine.

3.28. Localized viral disease:
1. is a major feature of congenital viral infections.
2. is not associated with a pronounced viremia.
3. can be associated with "carrier" individuals.
4. may have systemic clinical features, e.g., fever.

3.29. Clinical viral disease:
1. is most frequently due to toxin production.
2. is a poor indicator of virus infection.
3. can result without infection of host cells.
4. is associated with "target" organs in most disseminated infections.

3.30. Neutralizing antibodies:
1. may react with viral receptors.
2. are most effective at the time of infection or during viremia.
3. are the basis of serologic testing for viruses.
4. react with nucleocapsid antigens of enveloped viruses.

3.31. Antiviral nucleoside analogues:
1. are only effective against replicating viruses.
2. include acyclovir and AZT.
3. are frequently not administered systemically because they can be cytotoxic.
4. may block viral penetration.

Answers and Explanations

PRINCIPLES OF VIROLOGY

3.1. D. The period of time between the adsorption and penetration of the virus until the first appearance of intracellular virus is the eclipse phase.

3.2. D. Some persistent virus infections like serum hepatitis caused by hepatitis B virus have "carriers" who may not have clinical signs of the disease.

3.3. D. Interferon, a host cell glycoprotein produced in response to virus infection, is relatively host-specific but not virus-specific.

3.4. B. Nucleic acid hybridization is not a type of serologic test. ELISA tests are approximately 1000-fold more sensitive than the other serologic tests listed.

3.5. D. Since interferon is induced in the first cell infected by the virus, it is the first host defense mechanism that responds to a primary virus infection.

3.6. C. The viral genome codes for proteins and/or glycoproteins in both the capsid and the envelope.

3.7. B. The lipids present in the envelope make it dissociable by organic solvents.

3.8. C. The presence of an envelope and the characteristics of both the envelope and the capsid are important in viral classification.

3.9. D. Neither the envelope nor the capsid contains nucleic acid as an integral component, although nucleic acid is found inside the capsid.

3.10. B. The lipid portion of a viral envelope is usually derived from host cell membrane lipid.

3.11. A. Reassortment involves a one-for-one exchange of segments of nucleic acid and is therefore passed to succeeding viral generations. Transcapsidation involves no reciprocal genetic exchange.

3.12. B. Reassortment occurs with segmented RNA viruses, but transcapsidation can occur with either DNA or RNA viruses.

3.13. C. Because of the nature of the processes involved, only related viruses participate.

3.14. D. Neither process involves the DNA-ligase dependent reciprocal exchange of homologous regions of DNA, a characteristic of classic genetic recombination.

3.15. C. Both processes occur during assembly, not during nucleic acid replication as in classic genetic recombination.

3.16. C. ELISA tests use antiviral antibodies to detect viral antigens in tissues or cells and immunohistochemical staining uses antiviral antibodies to detect viral antigen in body fluids.

3.17. D. Viral inclusion bodies are observed in fixed and chemically stained tissues or cells. They are classified on the basis of their staining properties and morphology.

3.18. A. ELISA tests can be used to detect antiviral antibodies in serum.

3.19. C. Both ELISA tests for viral antigens and immunohistochemical staining depend on specific antiviral antibodies.

3.20. B. Impression slides made from imprints of infected tissue are used in place of tissue sections in some immunohistochemical staining procedures.

3.21. A. Live enveloped virus vaccine can employ attenuated virus strains that cause asymptomatic or mild clinical disease.

3.22. A. Live virus vaccines stimulate antibodies to all of the viral antigens since they replicate within the host. Killed virus vaccines do not replicate and only induce antibodies to the immunogen used.

119

3.23. C. Both live and subunit virus vaccines stimulate humoral immunity.

3.24. B. Since the killed vaccine contains no live virus, there is no possibility of a reversion to virulence, which is sometimes observed with live virus vaccines.

3.25. A. The virus present in live virus vaccines replicates within the host, thereby stimulating both cellular and humoral immunity.

3.26. B. The infectious virus particle is called a virion, not a viroid, and is inactivated by proteases, not by nucleases.

3.27. A. Amantadine inhibits the penetration and uncoating of influenza A virus.

3.28. C. Although localized infections are not associated with pronounced viremia, they can have clinical features associated with systemic viral infections accompanied by viremia.

3.29. C. Many viral infections are asymptomatic or subclinical. Clinical disease, however, is often associated with viral replication in "target organs" during disseminated infections.

3.30. A. Antibodies that neutralize the infectivity of enveloped viruses are directed against viral-specific envelope proteins or glycoproteins.

3.31. A. Nucleoside analogues inhibit viral replication by inhibiting viral DNA synthesis or function; they do not block penetration.

DNA Viruses

I. Human Adenoviruses

–are naked viruses with an icosahedral nucleocapsid composed of *hexons, pentons,* and *fibers.*

–have a *toxic activity* associated with pentons and *hemagglutinating activity* associated with pentons and fibers.

–contain *double-stranded DNA* that *replicates asymmetrically.*

–replicate in the nucleus of epithelial cells; the *E1A gene* is particularly important.

–produce localized infections of the eye, respiratory tract, gastrointestinal tract, and urinary bladder.

–produce subclinical infections frequently.

–are classified into seven groups (A through G) on the basis of their DNA homology.

A. Group A Adenoviruses

–do not cause a specific clinical disease in humans.

–induce *tumors in newborn hamsters.*

B. Group B Adenoviruses

–cause acute respiratory disease (ARD), pharyngoconjunctival fever, and hemorrhagic cystitis.

–can cause epidemics in military recruits (adenovirus Type 7).

C. Group C Adenoviruses

–cause about 5% of acute respiratory diseases in young children.

–cause latent infections in tonsils, adenoids, and other lymphatic tissue.

D. Group D Adenoviruses

–are associated with sporadic and epidemic keratoconjunctivitis.

–cause *"pink eye"* (adenovirus Type 8).

E. Group E Adenoviruses

–are associated with acute respiratory disease with fever and epidemic keratoconjunctivitis in military recruits.

F. Groups F and G Adenoviruses

–cause gastroenteritis.

G. Adenovirus Vaccines

–are used by the military for protection against Types 4 (Group E) and 7 (Group B) viruses.

–may be of the *subunit (hexons and fibers) type or live virus type.*

H. Diagnosis of Adenovirus Infections

–is most frequently done by observing a rise in neutralizing antibody titer.

–may be done by virus isolation from eyes, throat, or urine.

–may involve ELISA procedures on fecal specimens from gastrointestinal infections.

II. Hepadnaviruses

–have a complex virion structure that includes an envelope.

–have *circular, double-stranded DNA containing single-strand breaks,* a primer protein, and a viral specific DNA polymerase associated with the DNA.

–need a viral-coded reverse transcriptase to replicate since the virion DNA is synthesized from an RNA template.

–have their *DNA integrated into cellular DNA* during their replication.

–cause liver disease as indicated by their names: human hepatitis B virus, woodchuck hepatitis virus, and duck hepatitis virus.

A. Hepatitis B Virus (HBV)

–causes *serum hepatitis* and, frequently, subclinical infections.

–has a special name given to its virion—*the Dane particle.*

–is associated with the specific antigens: a surface antigen (*HB$_s$ag or Australian antigen)*, a core antigen (*HB$_c$ag*), and DNA polymerase-associated antigen (*HB$_e$ag*).

–has been implicated in hepatocellular carcinoma.

1. Serum Hepatitis

–has a long incubation period of 50 to 180 days.

–antigen-antibody complexes are involved in the disease process.

–is usually contracted through a parenteral route.

–can progress to a *chronic carrier state* with or without clinical symptoms.

–is a potential high risk for staff of hemodialysis units and health care workers exposed to fresh blood.

–may be determined to be an acute or chronic case on the basis of the amount of HB$_e$ag and HB$_s$ag and anti-HB$_s$ and anti-HB$_c$ by ELISA tests.

–may be treated prophylactically by (1) *passive immunization* with hepatitis B immune globulin (HBIG) or (2) *HB$_s$ag vaccine* produced from particles in healthy carriers or by recombinant DNA techniques.

III. Herpesviruses

–are enveloped viruses that have an icosahedral nucleocapsid containing *double-stranded DNA*.

–have linear DNA that consists of a short (18%) and a long (82%) component that are covalently linked and contain unique sequences flanked by inverted repeats.

–have a *tegument*, fibrous material between the nucleocapsid and envelope.

–replicate in the nucleus of the host cell.

–have *oncogenic potential*.

–have the ability to cause *latent infections* as well as acute infections.

–are classified into subfamilies: *Alphaherpesviruses* (herpes simplex viruses 1 and 2 and varicella-zoster virus), *Betaherpesviruses* (cytomegalovirus), and *Gammaherpesviruses* (Epstein-Barr virus).

A. Herpes Simplex Viruses (Types 1 and 2)

–have about 50% DNA sequence homology.

–produce both *common and type-specific antigens*.

–replicate by the regulated temporal synthesis of classes of proteins (α, β, and γ).

–produce a virus-specific DNA polymerase and thymidine kinase, which are necessary for replication.

–can cause transformation of hamster cells.

–are frequently *latent in neurons*.

–produce *defective virions* during serial passage.

–can produce (1) distinctive cytopathology (cell rounding and *polykaryocyte formation*) and/or (2) inclusion bodies (*Cowdry Type A inclusions*) in infected cells.

–are treated clinically by acyclovir, trifluridine, and vidarabine.

1. Herpes Virus Type 1 Disease

–may be a primary (gingivostomatitis) or recurrent (cold sores) infection.
–is usually clinically inapparent as a primary disease.
–usually presents as a lip, skin, or eye lesion.
–can progress to a severe fatal encephalitis.
–may be diagnosed by virus isolation or serologic testing involving neutralization or complement fixation tests or ELISA procedures.
–cannot presently be treated prophylactically by vaccine.

2. Herpes Simplex Virus Type 2 Disease

–may be a primary or recurrent infection.
–involves the genital area or the lip area.
–is most frequently sexually transmitted.
–includes *neonatal herpes*, which is a severe generalized disease of the newborn caused by virus infection during passage through an infected birth canal.
–may be diagnosed and treated as described for HSV-1.
–may include cervical and vulvar carcinoma since viral nonstructural antigens have been found in biopsies.

B. Varicella-Zoster Virus

–has a colchicine effect on human cells.
–cause an acute, primary disease (chickenpox) and a recurrent disease (zoster).
–is *latent in neurons*.

1. Varicella (Chickenpox)

–is a mild, highly infectious, generalized disease usually observed in children.
–is clinically characterized by vesicles on the skin and mucous membranes.
–can be diagnosed in stained smears or scrapings by observing multinucleated giant cells or viral antigens (fluorescent antibody staining) or by ELISA tests for antiviral antibodies.
– may be treated prophylatically in immunocompromised children by giving them *VZIG (varicella-zoster immune globulin)*.

2. Zoster

–is also called *shingles*.
–is a reactivated virus infection in adults.
–is characterized by severe pain and vesicles in a specific area of skin or mucosa supplied with nerves from one ganglia.
–can disseminate in immunocompromised individuals.

C. Cytomegalovirus (CMV)

–replicates more slowly and is more cell-associated than herpes simplex virus.
–replicates only in human fibroblasts.
–can transform hamster and human cells in vitro.
–causes an acute, primary infection and a latent infection that is only reactivated to clinical disease during immunosuppression.
–most commonly causes *inapparent disease* in children and adults, but can cause an *infectious mononucleosis-like disease*.
–causes *cytomegalic inclusion disease in infants*.
–forms *owls eye inclusions* in cells found in urinary sediments of infected individuals.

–can be isolated from the saliva and urine.

1. Cytomegalic Inclusion Disease

–is a generalized infection of infants with a distinct clinical syndrome that includes jaundice with hepatosplenomegaly, thromocytopenic purpura, pneumonitis, and central nervous system damage.

–is caused by intrauterine (congenital) or early postnatal infection.

–can cause fetal death.

–can be diagnosed by (1) virus isolation from urine or peripheral blood leukocytes, (2) serologic tests, and (3) new DNA hybridization tests involving the extraction of viral DNA from urine specimens.

D. Epstein-Barr Virus (EBV)

–infects human B lymphocytes.

–can transform human B lymphocytes.

–produces several distinct antigens, including a lymphocyte detected membrane antigen (*LYDMA*), a nuclear antigen (*EBNA*), an early antigen (*EA*), a membrane antigen (*MA*), and a viral capsid antigen (*VCA*).

–usually causes clinically inapparent infections but may cause *infectious mononucleosis* and is associated with Burkitt's lymphoma and nasopharyngeal carcinoma.

1. Infectious Mononucleosis

–is a disease of children and young adults (*kissing disease*) characterized by fever and enlarged lymph nodes and spleen.

–is also associated with the production of *atypical lymphocytes* and IgM *heterophile antibodies identified with the mononucleosis spot test*.

–can also be diagnosed by serologic tests involving indirect immunofluorescence procedures on fixed Epstein-Barr virus-producing cells or ELISA tests.

2. Burkitt's Lymphoma and Nasopharyngeal Carcinoma Patients

–have increased antibody titers to EBV virus.

–have cells that express EBNA and carry multiple copies of viral DNA.

IV. Parvoviruses

–are very small naked viruses with icosahedral nucleocapsids.

–contain *single-stranded DNA* and replicate in the nucleus.

–include *human parvovirus (HPV)* and *adenoassociated virus (AAV)*, a defective virus that requires adenovirus to replicate.

A. Human Parvovirus

–enters the body through the respiratory tract.

–infects and lyses progenitor erthyroid cells.

–may cause a febrile illness in blood recipients.

–can cause *aplastic crises in individuals with hemolytic anemias*.

–causes erythemic infectiosum (Fifth disease) in normal persons.

V. Papovaviruses

–are naked viruses with an icosahedral nucleocapsid that contains *double-stranded, circular DNA*.

–replicate in the nucleus of the cell.

–*produce latent and chronic infections* in their natural host.

–can induce tumors in some animals (see Oncogenic Viruses).

–include the *human papilloma (wart) virus, the JC virus* (isolated from patients with progressive multifocal leukoencephalopathy), and the *BK virus* (isolated from the urine of immunosuppressed patients).

–include the animal viruses: papilloma virus, polyma virus, and simian virus 40 (SV-40).

A. Human Wart Virus

–replicates in epithelial cells of the skin.

–is directly transferred from one person to another.

–causes *warts*.

–has been associated with benign cervical, vulvar, and penile cancers.

VI. Poxviruses

–have a complex brick-shaped virion that consists of an outer envelope that encloses (1) a core containing *linear double-stranded DNA* and (2) two *lateral bodies*.

–have more than 100 structural polypeptides, including many enzymes and a *transcriptional system associated with the virion*.

–replicate in the cytoplasm of the cell.

–are unique because *de novo formation of viral membrane* is required for replication.

–have *posttranslational cleavage* of proteins as part of their replication process.

–include (1) the human viruses: *vaccinia, variola, and molluscum contagiosum*; and (2) the animal viruses, which can cause highly localized (usually finger) occupational infections: *cowpox virus, paravaccinia virus (cows), and orf virus (sheep)*.

–produce eosinophilic inclusion bodies called *Guarnieri bodies* and membrane *hemagglutinins* in infected cells.

–share a common nucleoprotein (NP) antigen in their inner core.

–may be inhibited by rifampin (blocks envelope formation) and Marboran (interferes with late proteins and assembly).

A. Variola Virus

–causes *smallpox*, a generalized viral infection that has presumably been eradicated by a World Health Organization (WHO) vaccine program.

–grows on the chorioallantoic membrane of eggs, where it forms *pocks*, focal areas of viral-induced cellular necrosis.

–has LS antigens that form the basis for serologic tests.

–was treated prophylactically with Marboran and by vaccination with the vaccinia virus.

B. Vaccinia Virus

–is the variant of variola virus that generally produces only a mild disease and is used as *the immunogen in smallpox vaccination*.

–causes postvaccinal encephalitis in a minority (three per million) of vaccinated individuals.

–produces a 140-residue polypeptide that is closely related to epidermal growth factor.

–is being studied as a possible *immunizing vector* that would contain foreign genes for polypeptides, which would elicit neutralizing antibodies for other viruses, e.g., HSV-1 and 2.

C. Molluscum Contagiosum Virus

–infects epithelial cells where it causes a localized disease that resolves spontaneously in several months but may persist for 1 to 2 years.

–causes *small, wart-like lesions* on the face, arms, back, buttocks, and genitals.
–is transmitted by direct or indirect contact.
–can cause a sexually transmitted disease with papular lesions that can ulcerate and mimic genital herpes.
–forms characteristic *eosinophilic inclusion bodies* in infected cells.

Positive-Sense RNA Viruses

I. Coronaviruses

–are enveloped viruses that have a *helical nucleocapsid* that contains a *single-stranded RNA with positive (messenger) polarity.*
–have distinctive club-shaped surface projections that give the appearance of a *solar corona to the virion.*
–replicate in the cytoplasm and bind to cytoplasmic vesicles; therefore, there are no viral antigens on the surface of the infected cell.
–produce *colds* in adults (second most frequent cause) and are implicated in *infant gastroenteritis.*
–are not usually diagnosed in the laboratory, although a complement-fixation test and a neutralization test are available.
–are represented in humans by *prototype strains 229E and OC43.*
–are represented in mice by *mouse hepatitis virus,* which gives a chronic demyelinating disease that has been used as a model for multiple sclerosis in humans.

II. Picornaviruses

–are small naked viruses with an *icosahedral nucleocapsid* that contains a *single-stranded positive-sense RNA covalently linked to a small protein (VPg in poliovirus).*
–replicate in the cytoplasm of the cell, where their RNA is translated into a large polyprotein that is subsequently cleaved (*posttranslational cleavage*).
–modify their capsid proteins while they are in the assembled capsid.
–have cytopathic effects that include the formation of virus crystals in the cytoplasm of infected cells.
–are frequently *cytolytic* to infected cells.
–are classified into two groups: enterovirus and rhinovirus.

A. Enteroviruses

–cause a variety of human diseases that involve infections of the alimentary tract.
–are stable at acid pH (3-5).
–include the *polioviruses, coxsackie A and B viruses, echoviruses, enteroviruses, and hepatitis A virus.*
–isolates since 1969 are simply classified as enteroviruses and given a serotype number instead of being classified as either coxsackie viruses or echoviruses.

1. Poliovirus Infections

–are usually *subclinical* but can appear as (1) *mild illness,* (2) *asceptic meningitis,* or (3) an acute disease of the central nervous system (*poliomyelitis*) in which spinal cord motor neurons (anterior horn cells) are killed, and flaccid paralysis results.
–are caused by *three serotypes of virus.*
–can occur in epidemics.

–begin in the oropharynx and intestine but can travel inside axons to the spinal cord.
–are controlled by immunization with either a killed *vaccine (Salk)* or a live attenuated trivalent (contains the three serotypes) *vaccine (Sabin).*
–are diagnosed serologically by a complement-fixation test or by virus isolation from the throat (early in illness) or feces (later in illness).

2. Coxsackie Virus Infections

–are caused by *29 serotypes* of viruses that are divided into two groups, A and B, depending on the type of paralysis observed following inoculation into mice (A produces flaccid; B produces spastic).
–are associated with (1) *herpangina* (Type A); (2) *hand, foot, and mouth disease* (Type A); (3) *hemorrhagic conjunctivitis* (Type A); (4) *pleurodynia, myocarditis, pericarditis,* and meningoencephalitis (Type B); and (5) *aseptic meningitis and colds* (Types A and B).
–are the *most common cause of viral heart disease (Type B).*
–tend to occur in the summer and early fall.
–can be diagnosed by virus isolation from throat washings and/or stools or serologic tests using specific viral antigen if a particular virus strain is suspected.

3. Echovirus Infections

–are caused by over *30 serotypes* of viruses that initially infect the human enteric tract but cause diseases ranging from common colds and fevers, with or without rashes, to aseptic meningitis and acute hemorrhagic conjunctivitis.
–may be diagnosed by virus isolation from the throat or stools but is not done except *in summer outbreaks of aseptic meningitis or febrile illness with a rash.*

4. Enterovirus Infections

–are associated with various respiratory tract infections, central nervous system disease (enterovirus 71), and acute hemorrhagic conjunctivitis (enterovirus 70).

5. Hepatitis A Virus (HAV) Infections

–are caused by an enterovirus that could be called *enterovirus 72.*
–cause *infectious hepatitis.*
–are distinguished from hepatitis B virus infections because they have an abrupt onset and a relatively short incubation period (15 to 45 days) and are transferred by the fecal-oral route.
–are acute infections that can occur as epidemics.
–may be treated prophylactically by immune human globulin.
–are diagnosed serologically by increases in IgM using an ELISA test.

B. Rhinoviruses

–cause localized upper respiratory tract infections.
–are the most frequent cause of *the common cold.*
–exist in over *100 serotypes.*
–are acid-labile.

III. Retroviruses

–are enveloped viruses that probably have an *icosahedral capsid* that surrounds a *helical nucleocapsid, which contains an inverted dimer of linear single-stranded positive-sense RNA (diploid genome).*

–are divided into four types (A, B, C, and D) on a morphologic basis.

–are divided into three groups: *lentiviruses* (visna and maedi viruses of sheep and human immunodeficiency virus), *spumaviruses,* and *oncoviruses* (Types B, C, and D RNA tumor viruses).

–have a *virion-associated reverse transcriptase* (makes DNA copies from RNA) and replicate in the nucleus.

–need a host cell t-RNA to interact with reverse transcriptase (RT) before the RT-complex can bind to RNA and initiate DNA synthesis.

–have three distinctive genes: *gag (structural proteins), pol (reverse transcriptase), and env (envelope glycoproteins),* which are flanked by *long terminal repeat (LTR) sequences* that have regulatory functions.

–use posttranslation cleavage processes during the synthesis of gag and env gene products.

–cause mostly "slow" disease of animals or various cancers (see Oncogenic Viruses) with the exception of *human immunodeficiency virus (HIV)*, which causes acquired immune deficiency syndrome (AIDS).

A. Human Immunodeficiency Virus (HIV)

–has also been called human T-lymphotrophic virus Type III (HTLV-III).

–initiates infection by interaction of an envelope glycoprotein (gp-120) with the cellular T4 (CD4) lymphocyte surface receptor.

–synthesizes *transacting regulatory proteins, e.g., TAT protein*, which regulate viral transcription.

–infects and *kills helper T cells*, which depresses both humoral and cell-mediated immunity.

–travels throughout the body, particularly in macrophages.

–induces a distinctive cytopathogenic effect: *giant-cell (syncytia) formation.*

B. Acquired Immunodeficiency Syndrome (AIDS)

–results from suppression of the immune system.

–is preceded by an *AIDS-related complex syndrome (ARC).*

–is characterized by unusual cancers, e.g., Kaposi's sarcoma or severe opportunistic infections, e.g., *Pneumocystis carinii*, cytomegalovirus, and, frequently, an AIDS dementia complex (ADC) disease.

–is a high-risk disease for homosexuals, bisexual men, and IV drug users.

–does not always occur in individuals who have seroconverted.

–is diagnosed by clinical symptoms and serologic assays, including ELISA and Western Blot tests.

C. Endogenous Type C Retroviruses

–are viruses of Type C RNA tumor virus morphology (see Oncogenic Viruses).

–have a provirus that is a constant part of the genome of an organism and whose expression is regulated by the host cell.

–are genetically transmitted to all offspring.

–are not pathogenic for their hosts.

–are frequently induced to replicate when cells are placed in tissue culture.

–are called *ecotrophic* (multiply in cells of the species in which they were induced), *xenotrophic* (cannot infect cells from the species in which they were induced but infect other species), *amphotrophic* (grow in cells of the species from which they were induced as well as cells from other species).

IV. Togaviruses

–are enveloped viruses with an *icosahedral nucleocapsid containing single-stranded, positive-sense RNA*.

–have *hemagglutinins* associated with their envelope.

–replicate in the cytoplasm and posttranslationally process the polyproteins they synthesize.

–produce generalized infections.

–are divided in four groups of which three *(alphaviruses, flaviviruses, and rubiviruses)* are human pathogens.

A. Alphaviruses

–bud from the cell surface.

–have mosquito vectors and animal reservoirs.

–produce an encephalitis or a moderate systemic disease following the bite of a mosquito that has fed on an animal viral reservoir.

–lead to more serious encephalitis than flaviviruses.

–are diagnosed by serologic tests, usually complement fixation, because virus isolation is difficult.

–include *eastern equine encephalitis virus, western equine encephalitis virus*, and *Venezuelan equine encephalitis virus*.

B. Flaviviruses

–bud into the endoplasmic reticulum.

–produce three types of clinical symptoms: (1) an encephalitis (*St. Louis encephalitis virus*), (2) a severe systemic disease (*Yellow Fever virus*), and (3) a mild systemic disease characterized by a rash and muscle pain (*dengue*).

–have mosquito or tick vectors and wild bird or animal reservoirs.

–are diagnosed serologically except for Yellow Fever virus, which is diagnosed clinically.

1. Yellow Fever

–is a biphasic disease with clinical signs during initial virus replication in vascular endothelium and later replication in the liver.

–can be diagnosed by eosinophilic hyaline masses called *Councilman bodies* in the cytoplasm of infected liver cells.

–does not occur following immunization with the *attenuated vaccine strains, 17D and Dakar*.

2. Dengue

–is transmitted from monkey to humans by mosquitos.

–has characteristic skin lesions, as well as fever and muscle and joint pain.

–is sometimes called *break bone fever*.

–may be fatal if hemorrhages are associated with the infection.

C. Rubiviruses

–bud into the endoplasmic reticulum and the cell surface.

–include *rubella virus*.

1. Rubella Virus Infections

–cause *German measles*, a systemic infection characterized by lymphadenopathy and morbilliform rash.

–are often subclinical in adults.

–can produce *congenital infections*, which can cause serious damage to the infected fetus during the first 10 weeks of pregnancy and lead to rubella syndrome.

–are difficult to diagnose clinically, but can be diagnosed by serologic tests for IgM antibodies, including hemagglutination inhibition or ELISA.

–can be prevented by immunization with the *live attenuated vaccine strains HPV77 or RA 27/3.*

V. Other Viruses

A. Norwalk Virus

–is probably a member of the Calcivirus family.

–is a naked virus with an *icosahedral nucleocapsid containing a single-stranded, positive-sense RNA.*

–replicates in the cytoplasm.

–causes an *epidemic gastroenteritis.*

–has never been grown in tissue culture.

–may be demonstrated by a radioimmunoassay blocking test or immune adherence methods.

Negative-Sense RNA Viruses

I. Bunyaviruses

–are enveloped viruses with *three circular helical nucleocapsids*, each containing a unique piece of *single-stranded, negative polarity RNA* (the L, M, and S segments), viral nucleoprotein, and transcriptase enzyme.

–replicate with cytoplasm and *bud from the membranes of the Golgi apparatus.*

–can interact with serologically closely related virus to *produce recombinant viruses by genetic reassortment.*

–usually have rodent hosts and infect humans during an arthropod bite.

–produce mosquito-borne encephalitis (*California and LaCrosse encephalitis viruses*), sand fly and mosquito-borne fever (*Sandfly fever virus and Rift Valley fever virus*), and rodent-borne hemorrhagic fever (*Hantaan virus*).

A. Bunyavirus Encephalitis

–is the only bunyavirus disease endemic to the USA.

–is caused by the California and LaCrosse viruses, which occur mainly in the Mississippi and Ohio River Valley.

–has a small forest rodent reservoir for the virus and a mosquito vector.

–is usually mild (perhaps just meningitis) with excellent prognosis and rare sequelae.

II. Orthomyxoviruses

–are enveloped, spherical, or filamentous viruses with *eight helical nucleocapsids* containing a unique *single-stranded, negative-sense RNA.*

–have a *hemagglutinin* (H), a *neuraminidase* (N), and a *matrix protein* (M) associated with the envelope and a *transcriptase* (P), which is nucleocapsid-associated.

–form *defective interfering (DI) particles* that lack a segment of RNA necessary for productive replication.

–are assembled in the cytoplasm but depend on host nuclear functions, including RNA polymerase II, for transcription.

–are *influenza viruses* and classified as type A, B, or C, depending on a nucleocapsid antigen.

–have the capacity to undergo *genetic reassortment* due to the segmented nature of the genome.

–do not replicate well in tissue culture and are grown in animals or embryonated eggs.

–are designated by the nomenclature, which indicates virus type, species isolated from unless human, site of isolation, strain number, year of isolation, and hemagglutinin and neuraminidase subtype: for example, A/swine/New Jersey/8/76 (H_1N_1) and A/Phillippines/2/82 (H_3N_2).

A. Influenza Virus Hemagglutinin

–is an envelope glycoprotein that contains the *virus receptor* that binds to the cellular receptor site.

–agglutinates many species of red blood cells.

–induces neutralizing antibodies.

–has *fusion activity*, which allows the virion envelope to fuse with the host cell plasma membranes.

–is responsible for influenza epidemics when it antigenically changes.

–undergoes frequent minor mutations that result in antigenic changes that lead to *antigenic drift*.

–undergoes *antigenic shift* when major antigenic changes occur following reassortment between the hemagglutinin-coding RNA segments of animal or human viruses.

B. Influenza Virus Neuraminidase

–is an envelope glycoprotein that removes terminal sialic acid residues from oligosaccharide chains.

–is *involved in the release of virions* from infected cells.

–can undergo antigenic shift and drift mutations, but epidemics do not result from these changes.

C. Influenza

–is a localized infection of the respiratory tract.

–is usually not serious except in the elderly or in patients who suffer from a secondary bacterial pneumonia.

–is associated with Guillain-Barré syndrome (Types A and B) and Reye's syndrome (Type B).

–can be diagnosed by virus isolation or hemagglutination-inhibition serologic test.

–may be prophylactically treated with amantadine if Type A virus is involved or with a polyvalent killed vaccine containing the prevailing Types A and B strains.

III. Paramyxoviruses

–are spherical enveloped viruses with a *single helical nucleocapsid containing single-stranded, negative-sense RNA*.

–have a *hemagglutinin-neuraminidase (HN)*, a *fusion protein (F)*, and a *matrix protein (M)* associated with the envelope and a nucleocapsid associated *transcriptase (P)*.

–replicate in the cytoplasm of the cell.

–form *defective interfering particles (DI)* by segment deletions within the genome.

–form *heteroploid particles* (two nucleocapsids from unrelated paramyxoviruses in the same envelope) or *polyploid particles* (multiple copies of same nucleocapsid in a large envelope).

–cause acute and persistent infections.

–are divided into three genera on the basis of their chemical and biologic properties: paramyxoviruses (*parainfluenza and mumps viruses*), morbilliviruses (*measles virus*), and pneumoviruses (*respiratory syncytial virus*).

–exist in few antigenic types.

A. Paramyxovirus Hemagglutinin-Neuraminidase

–is a large surface glycoprotein with both *hemagglutinating* and *neuraminidase activity*, except in measles virus, which lacks neuraminidase activity and in respiratory syncytial virus where both activities have been lost.

–is *responsible for virus adsorption.*

–stimulates the production of neutralizing antibodies.

B. Paramyxovirus Fusion Protein

–is a surface glycoprotein with *fusion and hemolysin activities*, except in respiratory syncytial virus where the hemolysis activity is lost.

–is *responsible for virus penetration* into the cell.

–is composed of two subunits, F_1 and F_2, formed by proteolytic cleavage of precursor F_0 by a host enzyme.

C. Parainfluenza Virus Infections

–are caused by *parainfluenza 1 (Sendai virus).*

–cause a variety of upper and lower respiratory tract illnesses that usually occur in the fall and winter.

–cause *croup* (Type 2 virus) in infants.

–can be diagnosed using hemagglutination-inhibition of complement-fixation tests.

D. Mumps Virus Infections

–are frequently subclinical and occur in winter or early spring.

–result in a generalized disease characterized by the enlargement of one or both parotid glands.

–may affect testes and ovaries, causing swelling and pain.

–cause 10% to 15% of the cases of aseptic meningitis.

–are most frequently diagnosed by clinical observations, although the virus can be isolated from the saliva, cerebrospinal fluid, or urine.

–are inhibited by a *live attenuated vaccine* containing the Jeryl Lynn strain of virus. (It is usually included with live attenuated measles and rubella virus strains.)

E. Measles Virus Infections

–result in an acute generalized disease characterized by a maculopapular rash, fever, respiratory distress, and *Koplik's spots* on the buccal mucosa.

–are prevented by a *live attenuated measles vaccine* (Edmonston B or Schwartz strains) that is part of the trivalent vaccine (measles, mumps, and rubella) given to children.

–can produce *Warthin-Finkeldey cells* (large multinuclear cells) in nasal secretions.

–can progress to an *encephalomyelitis* (1:1000 cases) or *giant-cell pneumonia.*

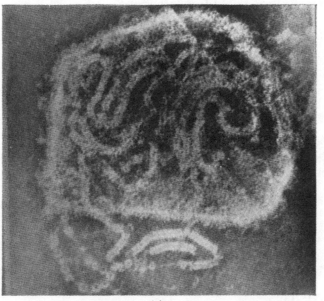

(a)

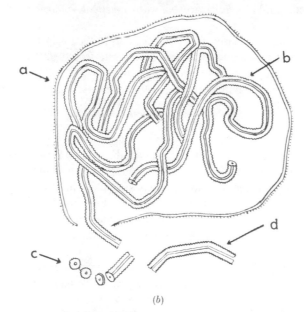

(b)

Figure 3.2. Mumps virus (M. parotidis) showing the internal helical structure (× 130,000). *a* = outer envelope, probably derived from host cell membrane (lipoprotein). *b* = internal nucleoprotein helical component. *c* = fragments of the helical component seen end on. Note hollow central region. *d* = helix released from the partially disrupted virus particle. (From Horne and Wildy, 1961.) (From Wilson GS, Miles A, eds. Topley and Wilson's Principles of Bacteriology, Virology and Immunity. 6th ed. Baltimore: Williams & Wilkins, 1975, vol 1, p 1232.)

—cause a temporary depression of cell-mediated immunity (due to viral infection of lymphocytes), which sometimes leads to secondary bacterial infections.

—can produce *subacute sclerosing panencephalitis,* a slow progressive degenerative neurologic disease of children and young adults.

F. Respiratory Syncytial Virus Infections

–are localized virus infections that are most often confined to the upper respiratory tract but can involve the lower respiratory tract.

–are the *major cause of serious bronchiolitis and pneumonia in infants.*

–may have an immediate hypersensitivity component.

–are caused by an extremely labile virus that produces a characteristic syncytial effect (cell fusion) in infected cells.

–may be rapidly diagnosed by a direct immunofluorescent test on exfoliated cells in nasopharyngeal smears.

–may be serologically diagnosed by ELISA test in adults, but infants seldom develop antibodies.

G. Newcastle Disease Virus Infections

–are caused by a paramyxovirus that is a natural respiratory tract pathogen of birds, particularly chickens.

–occur as *an occupational disease* of poultry workers.

–are observed clinically as a *mild conjunctivitis without corneal involvement.*

IV. Rhabdoviruses

–are enveloped, *bullet-shaped* viruses with a *helical nucleocapsid containing single-stranded, negative-sense RNA.*

–they have a virion-associated transcriptase and replicate in the cytoplasm.

–generate deletion mutants that *form defective interfering (DI) particles.*

–establish persistent infections in cell cultures.

–are represented by the human pathogen, *rabies virus,* and the bovine pathogen, *vesicular stomatitis virus.*

A. Rabies Virus

–produces specific cytoplasmic inclusion bodies called *Negri bodies* in infected cells.

–can travel throughout the nervous system in nerve fibers.

–has a predilection for the hippocampus (Amnon's horn cells).

–is called *street virus* if freshly isolated and *fixed virus* if serially passaged in rabbit brain so that it no longer multiplies in extraneural tissue.

–produces disease following inoculation by an animal bite and, occasionally, by inhalation.

–produces a fatal disease unless the infected individual received previous immunization or postexposure prophylaxis consisting of passive immunization with human rabies immune globulin and immunization with a vaccine.

–is identified in suspected tissues by a direct immunofluorescence test.

–is grown in rabbit brain (Semple's vaccine), embryonated eggs (*duck embryo vaccine*), or W1-38 cells (*human diploid cell vaccine*) before inactivation and use as a vaccine.

–has been attenuated by growth in chick embryo for use as an animal, not human, vaccine (*Flury's vaccine*).

B. Vesicular Stomatitis Virus

–causes foot and mouth disease in cattle.

–is well known for its ability to produce DI particles and persistent infections.

V. Other Viruses

A. Marburg and Ebola Viruses

–belong to a proposed virus family called *Filovirus*.
–are enveloped viruses with *a helical nucleocapsid containing single-stranded, negative-sense RNA.*
–produce *African hemorrhagic fevers* that often lead to death.

Other RNA Viruses

I. Arenaviruses

–are enveloped viruses with *two string-of-beads nucleocapsids,* each containing a unique *single-stranded, circular RNA.*
–have one molecule of genomic RNA (the L or large) that is of negative polarity and one molecule (the S or short) that is *ambisense,* i.e., has both negative and positive sense.
–replicate in the cytoplasm and have *host cell ribosomes* in their virion.
–infect mice and/or rats as their natural hosts.
–are initially passed from rodents to humans but can then be transferred by direct human contact.
–cause some highly contagious hemorrhagic fevers (*Junin, Machupo, and Lassa viruses*) that are not endemic to the U.S. and a meningitis or flu-like illness (lymphocytic choriomeningitis virus) that is endemic.

II. Reoviruses

–are naked viruses with a *double-shelled (outer shell and core) icosahedral capsid containing 10 or 11 segments of double-stranded RNA.*
–replicate in the cytoplasm.
–have a core-associated transcriptase.
–are classified into three groups: reoviruses, rotaviruses, and orbiviruses.

A. Reoviruses

–have ten segments of double-stranded RNA.
–have an *outer shell-associated hemagglutinin (σ 1)* that (1) agglutinates human 0 or bovine erythrocytes, (2) is the *viral receptor* and therefore determines tissue trophism, and (3) is the determinant for the three serotypes of reoviruses.
–form distinctive eosinophilic inclusion bodies.
–*replicate their RNA conservatively*, not semiconservatively.
–produce *minor upper respiratory tract infections and gastrointestinal disease*, but are frequently recovered from healthy people also.
–are diagnosed by a complement-fixation test and serotyped by hemagglutination-inhibition assays.
–may be isolated from feces and throat washings.

B. Rotaviruses

–have *11 segments of double-stranded RNA.*
–exist in at least four serotypes, with Type A being responsible for most human infections.
–cause *infantile diarrhea* and are the most common cause of *gastroenteritis in children.*
–are frequent causes of nosocomial infections.

–are diagnosed by demonstrating virus in the stool or serologic tests, particularly ELISA.

C. Orbiviruses

–have *ten segments of double-stranded RNA.*
–infect insects, which transfer the virus to humans.
–cause mild fevers in humans.
–are represented by *Colorado tick fever virus,* which is carried by the wood tick, *Dermacentor andersoni.*

III. Miscellaneous

A. Delta (δ) Agent

–is a virus that has *circular, single-stranded RNA molecules and an internal core δ antigen surrounded by a hepatitis B virus envelope.*
–*is defective* and can only replicate in the presence of hepatitis B virus.
–is associated with both acute and chronic hepatitis and always with hepatitis B virus.
–causes a more severe hepatitis than hepatitis B virus alone.
–may be diagnosed serologically with an ELISA test.

B. Non-A, Non-B Hepatitis Viruses

–exist in two types: (1) a naked virus with an *RNA genome* and (2) an enveloped virus that may be a retrovirus or have a *DNA genome.*
–infects the body parenterally and after a 14 to 20 day incubation period causes hepatitis.
–causes *90% of blood transfusion-associated or blood product administration-associated hepatitis.*
–can cause *chronic infections* and, therefore, individuals who are *carriers.*

Slow Viruses and Prions

I. Subacute Sclerosing Panencephalitis Virus

–is a variant or very close relative of measles virus.
–*causes subacute sclerosing panencephalitis (SSPE),* a rare, fatal, slowly progressive, demyelinating central nervous system disease of teenagers and young adults.
–may result from improper synthesis or processing of the matrix (M) viral protein.

II. JC Virus

–is a papovavirus that infects humans frequently, but rarely produces disease unless the host is immunosuppressed.
–has been isolated from patients with *progressive multifocal leukoencephalopathy (PML),* a rare central nervous system disease.
–causes demyelination by infecting and killing oligodendrocytes.

III. Animal Lentiviruses

–are retroviruses that cause slow generalized infections of sheep (visna and progressive pneumonia virus) and goats (caprine arthritis virus).
–produce minimal amounts of infectious virus in their hosts.
–undergo *considerable antigenic variation* in their host due to mutations in envelope glycoproteins.

IV. Prions

–are not viruses, but are proteinaceous material lacking nucleic acid.

–are associated with four degenerative central nervous sytem diseases (*the subacute spongiform virus encephalopathies*): kuru and Creutzfeldt-Jakob disease of humans, scrapie of sheep, and transmissible encephalopathy of mink.

–have a *prion protein (PrP)* that is associated with their infectivity, yet is encoded by a cellular gene.

Oncogenic Viruses

I. Oncogenic Viruses

–are classified as DNA or RNA tumor viruses.

–produce tumors when they infect appropriate animals.

–transform (alter cell growth, cell surface antigens, and biochemical processes) infected cells.

–introduce "transforming genes" or induce expression of quiescent cellular genes, which results in the synthesis of a transforming protein/s.

II. DNA Tumor Viruses

–cause *transformation in nonpermissive cells* (cells that are infected, but do not support total virus replication).

–form *proviruses*.

–include human papillomaviruses: adenoviruses, hepatitis B virus, Epstein-Barr virus, molluscum contagiosum virus, JC and BK viruses, and, maybe, HSV-2.

–include the animal viruses: chicken *Marek's disease virus* (a herpesvirus), mouse *polyomavirus* (a papovavirus), and monkey *SV-40 virus* (a papovavirus).

A. SV-40 Virus

–undergoes productive replication in monkey cells but transforms nonpermissive hamster and mouse cells.

–synthesizes an early protein called *large T (tumor) antigen,* which associates with a cellular phosphoprotein (P53), binds to DNA, has ATPase activity, initiates viral DNA replication, and *establishes and maintains SV-40 induced transformation.*

–synthesizes two other tumor antigens, middle T and small T.

B. Polyomavirus

–grows permissively in mouse cells but transforms nonpermissive hamster and rat cells.

–*synthesizes a transforming large T antigen.*

III. RNA Tumor Viruses

–are also called *oncornaviruses.*

–are *retroviruses* (Oncovirus group).

–infect permissive cells but transform rather than kill.

–cause tumors of the reticuloendothelial and hematopoietic systems (leukemias), connective tissues (sarcomas), or mammary gland.

A. Type B Tumor Viruses

–have an eccentric electron-dense core structure in their virion.

–are best exemplified by *mouse mammary tumor virus (MMTV), the Bittner virus.*

B. Type C Tumor Viruses

–have electron-dense cores in the center of the virion.

–include most of the RNA tumor viruses.

–exist in two classes on the basis of their replicative ability: nondefective and defective.

–contain a cellular-derived *oncogene* (codes for a cancer-inducing product) as well as *virogenes* (gag, pol, and env), except for a few nondefective murine leukosis viruses (AKR and Moloney) that have no oncogenes.

1. Oncogenes

–are genes that cause cancer.

–have copies in viruses (*v-onc*) and cells (*c-onc or protooncogene*).

–are "switched off" or down-regulated in normal cells.

–have products that are essential to normal cell function or development.

–may code for proteins, which can be (1) *tyrosine protein kinases* (src gene-Rous sarcoma virus, abl gene-Abelson leukemia virus), (2) *guanine-nucleotide binding proteins* (Ha-ras-Harvey sarcoma virus), and (3) *chromatin-binding proteins* (myc-MC29 myclocytomatosis virus and fos-FBJ osteosarcoma virus).

–may code for *cellular surface receptors*, e.g., epidermal growth factor receptor (erb-B product of avian erythrobastosis virus).

–may code for *cellular growth factors*, e.g., platelet-derived growth factor (SIS product of simian sarcoma virus).

2. Nondefective Viruses

–have all their virogenes and can therefore replicate themselves.

–have high oncogenic potential if they also contain an oncogene, e.g., *Rous chicken sarcoma virus.*

–have low oncogenic potential if they do not have an oncogene, e.g, *AKR and Moloney murine leukemia viruses, and human T-cell leukemia viruses I and II.*

3. Defective Viruses

–have a virogene or part of a virogene replaced by an oncogene.

–need *helper viruses* to provide missing virogene product for replication.

–have high oncogenic potential, e.g., murine sarcoma viruses (*Kirsten and Harvey viruses*) and murine leukemia viruses (*Friend, Abelson, etc.*).

4. Human T-Cell Leukemia Viruses

–are nondefective exogenous retroviruses.

–replicate and transform T4 antigen-positive cells.

–produce giant multinucleated cells.

–have no identifiable oncogene.

–cause *adult T-cell leukemia (ATL).*

Review Test

HUMAN VIRUSES

DIRECTIONS: For each numbered item, select one lettered heading that is most closely associated with it. Each lettered heading may be selected once, more than once, or not at all.

A. Varicella-zoster virus
B. Coronavirus
C. Epstein-Barr virus
D. Parvovirus
E. Papovavirus

3.1. Contains single-stranded DNA. *D*
3.2. Stimulates anti-LYDMA antibodies. *C*
3.3. Contains circular, double-stranded DNA. *E*
3.4. Is latent in human neurons. *A*
3.5. Has a genome of positive sense. *B*

A. *coronavirus*
B. Flavivirus
C. Echovirus
D. Newcastle disease virus
E. Rotavirus

3.11. Aseptic meningitis. *C*
3.12. Colds. *A*
3.13. Conjunctivitis. *D*
3.14. Infantile diarrhea. *E*
3.15. Encephalitis. *B*

A. Australian antigen
B. Cowdry Type A inclusions
C. Owl's eye inclusions
D. Koplik's spots
E. Heterophile antibodies *(mono)*

3.6. Cytomegalovirus infection. *C*
3.7. Herpes simplex virus infection. *B*
3.8. Epstein-Barr virus infection. *E*
3.9. Measles virus infection. *D*
3.10. Hepatitis B virus infection. *A*

A. Coronavirus

A. Dane particle
B. Infectious hepatitis
C. Double-shelled capsid
D. Defective virus
E. Diploid genome

3.16. δ agent. *D*
3.17. Hepatitis B virus. *A*
3.18. Hepatitis A virus. *B*
3.19. Human immunodeficiency virus. *E*
3.20. Reovirus. *C*

DIRECTIONS: Each set of lettered headings below is followed by a list of numbered words or phrases. Choose answer.

 A. if the item is associated with A only.
 B. if the item is associated with B only.
 C. if the item is associated with both A and B.
 D. if the item is associated with neither A nor B.

A. Respiratory syncytial virus
B. Parainfluenza virus
C. Both
D. Neither

3.21. Causes respiratory tract infections of infants. *RSV + paraninf.*

3.22. Induces characteristic giant, multinucleated cells. *RSV*
3.23. Has a hemagglutinin glycoprotein that is distinct from a neuraminidase glycoprotein. *Neither*
3.24. Causes serious upper respiratory tract infections in adults. *neither*
3.25. Causes croup. *Type 2 parainfluenza*

139

A. Bunyaviruses
B. Arenaviruses
C. Both
D. Neither
3.26. Contain a DNA genome. *Neither, both have single DNA*
3.27. Have a segmented "ambisense" genome. *B*
3.28. Have arthropod vectors. *A*
3.29. Produce gastroenteritis. *Neither*
3.30. Can cause an encephalitis. *A*

A. AIDS
B. Serum heptatitis
C. Both
D. Neither
3.31. Is caused by a virus that uses a reverse transcriptase during replication. *AIDS+HEP B*
3.32. Is a high-risk disease for IV drug users. *Both*
3.33. Results from a suppression of the immune system. *A*
3.34. Can be diagnosed by ELISA tests. *Both*
3.35. Can be associated with the delta (δ) agent. *B*

A. Measles
B. German measles (Rubella)
C. Both
D. Neither
3.36. May be prevented by immunization with a live-attenuated vaccine. *C*
3.37. Is associated with congenital infections. *B*
3.38. Is caused by a positive-sense RNA virus. *B*
3.39. Is associated with antihemagglutinin antibodies. *Both*

3.40. May lead to a slowly progressive, demyelinating CNS disease. *measles*

A. SV-40 virus
B. Rous sarcoma virus
C. Both
D. Neither
3.41. Form proviruses. *Both*
3.42. Contain oncogenes. *B*
3.43. Are only able to transform permissive cells. *B*
3.44. Synthesize transforming viral-specific proteins that bind to cellular DNA. *A*
3.45. Are defective viruses that need a helper virus to replicate. *D*

A. Varicella-zoster virus
B. Cytomegalovirus
C. Both
D. Neither
3.46. Is frequently reactivated during immunosuppression. *Both*
3.47. Causes a disease prevented by a live, attenuated vaccine. *no vaccines B*
3.48. Causes a serious congenital infection. *B*
3.49. Causes a disease that may be prevented by passive immunization with a product available from the Red Cross. *A*
3.50. Can be isolated from the urine. *B*

DIRECTIONS: For each of the questions or incomplete statements below, *one* or *more* of the answers or completions given is correct. Choose answer:

A. if only **1, 2,** and **3** are correct
B. if only **1** and **3** are correct
C. if **2** and **4** are correct
D. if only **4** is correct
E. if all are correct

3.51. Diseases associated with human adenoviruses include
1. acute respiratory disease.
2. "pink eye."
3. pharyngoconjunctival fever.
4. tumors in newborn hamsters.

3.52. Viral hepatitis
1. may be caused by a single-stranded DNA virus.
2. may be diagnosed by an ELISA test.
3. is an example of a latent infection.
4. may be caused by several distinct types of viruses.

3.53. Rabies
1. is caused by a rhabdovirus.
2. may be prevented by passive immunization.
3. may be diagnosed by characteristic inclusions called "Negri bodies" in infected cells.
4. has a predilection for the Amnon's horn cells of the hippocampus.

3.54. Influenza A virus
1. causes localized infections.
2. has the capacity to undergo genetic reassortment.
3. undergoes antigenic drift.
4. has a nonsegmented, negative-sense genome.

3.55. Endogenous Type C viruses

1. are defective viruses.
2. are not pathogenic for their hosts.
3. cause tumors in their hosts.
4. have a provirus form.

3.56. The skin rashes associated with virus infections

1. are the result of virus replication in the skin.
2. are found with localized virus infections.

3. can aid in the diagnosis of virus infections.
4. are due to viremia.

3.57. Viral gastroenteritis

1. is caused by nonenveloped viruses.
2. is associated with RNA viruses.
3. frequently occurs in epidemics.
4. is often diagnosed by ELISA tests on fecal samples.

Answers and Explanations

HUMAN VIRUSES

3.1. D. Parvoviruses are very small naked viruses with a single-stranded DNA genome.

3.2. C. Epstein-Barr virus produces a lyphocyte detected membrane antigen (LYDMA) early during active replication and in nonvirus-producing B lymphocytes cell lines transformed by the virus.

3.3. E. Papovavirus, naked viruses with a double-stranded circular DNA genome, can induce tumors in some animals.

3.4. A. Varicella-zoster virus, which can become latent in neurons within the brain, is sometimes reactivated to cause shingles.

3.5. B. Coronaviruses are enveloped viruses with a genome consisting of a single positive-sense RNA molecule.

3.6. C. Cells with "owl's eye" inclusions are frequently found in the urine of individuals with cytomegalovirus infections.

3.7. B. Herpes simplex virus frequently produces eosinophilic nuclear inclusion bodies classified as Cowdry Type A in infected cells.

3.8. E. Infection with Epstein-Barr virus can cause infectious mononucleosis, which may be characterized by the appearance of heterophile antibodies.

3.9. D. Koplik's spots (exanthems of the buccal mucosa) appear during the prodromal period of measles.

3.10. A. The hepatitis B virus surface antigen is also called the Australian antigen.

3.11. C. Echoviruses are a type of picornavirus that can cause summer epidemics of aseptic meningitis.

3.12. A. In addition to rhinoviruses, coronaviruses are frequently associated with "colds" in adults.

3.13. D. Newcastle disease virus is a respiratory tract pathogen of chickens that can be transferred to poultry workers and causes a mild conjunctivitis.

3.14. E. Rotaviruses, a member of the reovirus family, cause a major portion of infantile diarrhea.

3.15. B. St. Louis encephalitis virus, a flavivirus with a mosquito vector, causes an important arboviral encephalitic disease in the U.S.

3.16. D. The δ agent is a defective virus with a single-stranded RNA genome inside a hepatitis B virus envelope.

3.17. A. The 42nm hepatitis B virus virion is also called the Dane particle.

3.18. B. Infectious hepatitis is caused by hepatitis A virus, a member of the picornavirus family.

3.19. E. Human immunodeficiency virus has a genome that contains two linked copies of single-stranded, positive-sense RNA.

3.20. C. Reoviruses have a unique double-stranded capsid that contains a segmented double-stranded RNA genome.

3.21. C. Both viruses cause upper respiratory tract infections in infants.

3.22. A. As indicated by its name, respiratory syncytial virus causes the formation of characteristic giant cells that can be observed in nasal secretions.

3.23. D. Both of these viruses are members of the paramyxovirus family, which has a single envelope glycoprotein with both activities. Orthomyxoviruses have separate glycoproteins.

142

3.24. D. Neither virus causes serious upper respiratory tract infections in adults, although both may do so in infants.

3.25. B. Croup, an early childhood upper respiratory tract infection, is caused by Type 2 parainfluenza viruses.

3.26. D. Both bunyaviruses and arenaviruses have a single-stranded RNA genome.

3.27. B. The short genomic RNA molecule of arenavirus is "ambisense," the 3 foot half having negative sense and the 5 foot half having positive sense.

3.28. A. The California and LaCrosse encephalitis viruses, which have a mosquito vector, are bunyaviruses.

3.29. D. Viruses within these families are not involved in human gastroenteritis.

3.30. A. Two of the arboviral encephalitis viruses, California and LaCrosse, are bunyaviruses.

3.31. C. Both the human immunodeficiency virus, which causes AIDS, and the hepatitis B virus, which causes serum hepatitis, use a virus-coded reverse transcriptase during replication.

3.32. C. Since the virus that causes both diseases is present in the blood, both diseases are of high-risk for IV drug users.

3.33. A. The virus that causes AIDS depletes the "helper" T lymphocyte population, which diminishes the immune capabilities of the infected individual.

3.34. C. There are excellent ELISA tests for the various viral antigens and antibodies associated with these diseases.

3.35. B. The δ agent, a defective virus, causes a more severe form of serum hepatitis than that observed with hepatitis B virus alone.

3.36. C. The Edmonton strain of measles virus and the HPV72 strain of rubella virus are used in live attenuated vaccines.

3.37. B. Rubella virus can cause a congenital rubella syndrome if the fetus is infected during the first 10 weeks of pregnancy.

3.38. B. Rubella virus is a togavirus and therefore has a single-stranded, positive-sense RNA;

measles virus is a paramyxovirus that has negative-sense, single-stranded RNA.

3.39. C. Both rubella and measles viruses have enveloped-associated glycoproteins with hemagglutinating activity.

3.40. A. Persistent infection by some measles viruses can lead to a slowly progressive, demyelinating CNS disease called subacute sclerosing panencephalitis.

3.41. C. Both of these viruses synthesize a double-stranded DNA that can integrate into a host cell chromosome.

3.42. B. Rous sarcoma virus is a retrovirus that contains the src oncogene.

3.43. B. SV-40 virus will replicate in permissive cells and kill them; Rous sarcoma virus will replicate and transform permissive cells.

3.44. A. The large T antigen synthesized by SV-40 virus binds to cellular DNA; the src protein synthesized by Rous sarcoma virus is a tyrosine protein kinase.

3.45. D. Both viruses are capable of independent replication in permissive cells.

3.46. C. Immunosuppression can cause the reactivation of varicella-zoster virus from neurons and cytomegalovirus from various sites in the body.

3.47. D. No vaccines are available for either of these viruses.

3.48. B. Intrauterine infection with cytomegalovirus can cause a serious infection of the newborn, called cytomegalic inclusion disease.

3.49. A. Primary infection of immunocompromised children by varicella-zoster virus can be prevented by prophylactic treatment with VZIG (varicella-zoster immune globulin).

3.50. B. Cytomegalovirus can frequently be isolated from the urine.

3.51. E. Various types of human adenoviruses are associated with all of the diseases listed.

3.52. C. Several viruses, including hepatitis A and B viruses, non-A, non-B viruses, the delta agent, yellow fever virus, and others, cause viral hepatitis. ELISA tests are available for the diagnosis of hepatitis caused by hepatitis A, B, and δ agent viruses.

3.53. E. All of the statements listed concerning rabies are true.

3.54. A. Influenza A virus has a segmented genome composed of eight pieces of negative-sense, single-stranded RNA.

3.55. C. Endogenous Type C viruses are retroviruses that are not pathogenic for their host and often replicate (therefore are nondefective) when cells harboring them are placed in culture.

3.56. B. Skin rashes are frequently associated with generalized or systemic viral infections and result from viral replication in the cells that compose the skin.

3.57. E. All of the statements concerning viral gastroenteritis are true.

4

The Fungi

General Characteristics

I. Fungi

–are eukaryotic.
–are commonly called yeasts, molds, and mushrooms.
–are heterotrophic and absorptive in metabolism.
–have a complex cell wall.
–belong to a kingdom, the Fungi, separate from the plants and bacteria.
–are subdivided into four medically important phyla: Zygomycota, Ascomycota, Basidiomycota, and Deuteromycota.
–reproduce typically by asexual and sexual mechanisms.

II. The Fungal Thallus

–is the body of the fungus.
–may consist of single, round to oval cells (called yeasts); *or*
–may consist of hyphae or filamentous structures with or without cross walls or septae (the molds and mushrooms); *or*
–may convert from a hyphal or spore form to a yeast form (and vice versa) depending on environmental conditions in those fungi called the dimorphic fungi (most notably: *Blastomyces, Histoplasma, Coccidioides,* and *Sporothrix*).

III. The Fungal Cell Wall

–protects cells from osmotic shock and determines shape.
–is a multilayered, fibrillar structure that is refractile under light microscopy.
–is composed primarily of polysaccharides, most notably chitin but also glucans and mannans.
–has proteins associated with it.
–is antigenic.
–is stained with the Periodic Acid Schiff Reaction and the Methenamine Silver Stain.

IV. The Fungal Cell Membrane

–has a typical eukaryotic bilayered membrane.
–is unique in that the major sterols are ergosterol and zymosterol, unlike the human cell membrane, which has cholesterol as the primary sterol.

V. Fungal Cellular Components

–include eukaryotic nuclei, mitochondria, and numerous vacuoles.

–do not include chloroplasts.

VI. Hyphal Forms

–may be septate (with cross walls) or aseptate (without regularly occurring cross walls).

–may be dematiaceous (dark colored: typically olive brown to black) or hyaline (colorless).

–are multinuclear with cytoplasmic streaming occurring across septae spores between the cells.

–grow apically.

–in a mass are referred to as a mycelium.

–in a more organized body with spores are referred to as a fruiting body.

VII. Yeast Forms

–are single-celled fungi, generally round to oval.

–generally have thicker cell walls.

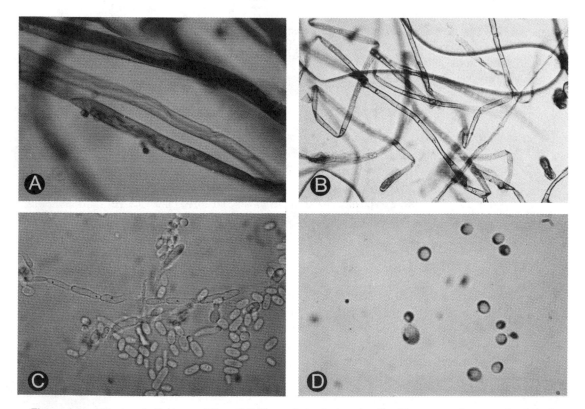

Figure 4.1. Microscopic Features of Fungal Cultures. **A,** Lactophenol-aniline blue preparation revealing broad, ribbon-like, aseptate hyphae characteristic of the Zygomycota. **B,** Septate hyphae of a hyaline mold. **C,** Pseudohyphae of a yeast colony resembling links of sausages in a chain. Pseudohyphae are differentiated from true hyphae, which are usually considerably longer and have parallel walls with no indentations at the points of septation. **D,** Appearance of budding yeast cells (blastoconidia) as they appear in direct mounts prepared from yeast culture. (From Koneman EW, Roberts GD. Practical Laboratory Mycology. 3rd ed. Baltimore: Williams & Wilkins, 1985, pp 69, 71.)

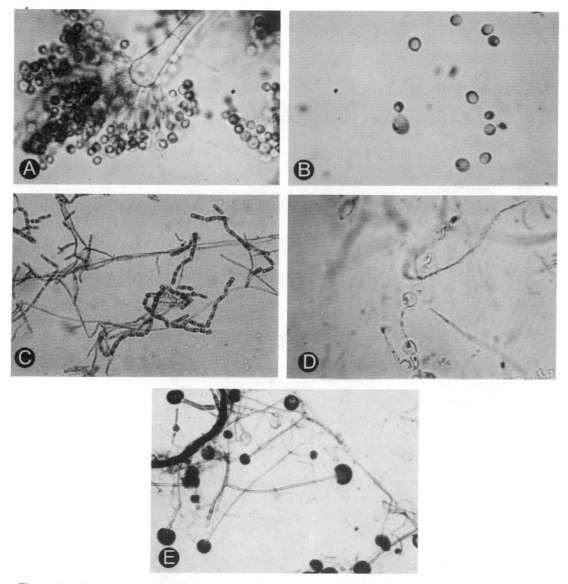

Figure 4.2. Structures of conidial sporulation. **A,** Fruiting head of *Aspergillus* sp demonstrating a club-shaped vesicle, phialides radiating from the top portion of the vesicle and chains of conidia. High power. **B,** Microscopic features of fungal cultures. Appearance of budding yeast cells (blastoconidia) as they appear in direct mounts prepared from yeast cultures. **C,** Appearance of arthroconidia as they typically appear in lactophenol-aniline blue mounts. **D,** Intercalary chlamydoconidia most often observed in old fungal cultures. **E,** Photomicrograph of a lactophenol-aniline blue mount of a Zygomycete illustrating dark-staining sporangia supported by long sporangiospores. Low power. (From Koneman EW, Roberts GD. Practical Laboratory Mycology. 3rd ed. Baltimore: Williams & Wilkins, 1985, pp 71, 73.)

–may also be multinuclear.
–generally reproduce by budding (blastoconidia).
–may have a capsule (most notably *Cryptococcus*).

VIII. Dimorphic Fungi

–are capable of converting from a yeast or yeast-like form to a filamentous form and vice versa.
–are stimulated to convert by environmental conditions.
–usually exist in the yeast or yeast-like form in a mammalian body.
–usually exist as the filamentous form in the environment (i.e., *Coccidioides* hyphae and arthroconidia in the desert soil).
–are most notably *Blastomyces, Histoplasma, Coccidioides,* and *Sporothrix* in the U.S.

IX. Pseudohyphae

–are a series of elongated blastoconidia remaining attached to each other, forming a hyphal-like structure but with constrictions at the septations.
–are characteristic of *Candida* species.

X. Fungal Spores or Propagules

–may be formed either asexually (without nuclear fusion) or by a sexual process (involving nuclear fusion and then meiosis).

A. Asexual Fungal Propagules Include

–**conidia**, which are asexual spores formed on the outside of a specialized fruiting structure called a conidiophore.
–**blastoconidia**, which are buds from a mother yeast cell.
–**arthroconidia**, which are conidia formed by the fragmentation of the hyphal strand.
–**chlamydoconidia**, which are thick-walled resting spores, generally spherical in shape and occurring terminally or within the hyphae.
–**sporangiospores**, which are spores formed inside of a specialized fruiting structure called a sporangium.

B. Sexual Spores Include

–**zygospores**, which are thick-walled zygotes formed by fusion of two hyphal tips.
–**ascospores**, which occur in multiples of four inside a structure called an ascus.
–**basidiospores**, which are sexual spores (haploid) formed on the outside of a structure called a basidium.

XI. The Classification of Fungi

–the characteristics of four medically important phyla are described in the following table:

Taxonomy of the Fungi Involved in Infections

Phylla [with old Class names in ()]	Habitat	Basic Units	Sexual Spores	Representative Organisms Causing Human Disease
Zygomycota (Phycomycetes)	Soil, bread, compromised hosts	**Aseptate hyphae**	**Zygospores** & a few others	***Rhizopus, Absidia, Mucor*** (Major ones causing infection)
Basidiomycota (Basidiomycetes)	Soil & wood (mushrooms); grains (ergot); humans less commonly	**Septate hyphae** or **yeasts**	**Basidiospores**	**Cryptococcus**
Ascomycota (Ascomycetes)	Fruits, organic material, and people	**Septate hyphae** or **yeasts**	**Ascospores**	Many
Deuteromycota (**Fungi Imperfecti**) (Deuteromycetes)	Soil, animals	**Septate hyphae** or **yeasts**	**None**	Most of the pathogens

Fungal Diseases

I. Medical Problems

–allergy.
–"poisoning."
–infection.

II. Fungal Allergies

–are common since molds will grow on any damp organic surface and spores are constantly in the air.
–generally occur in individuals with other allergies.

III. Fungal "Poisonings"

–include mycotoxicosis (ingestion of mold-contaminated food), which is primarily an animal problem except for aflatoxin (produced in moldy peanuts), which is a known carcinogen.
–include mycetismus (mushroom poisoning), which varies from mild GI tract disturbances or mild neurologic disturbances to very severe poisoning, which may be fatal.

IV. Fungal Infections

–are called mycoses (singular: mycosis).
–range from superficial to overwhelming infections, rapidly fatal in the compromised host.
–are increasing in frequency due to the increased usage of antibiotics, corticosteroid, and cytotoxic drugs.
–in a severely compromised host may be caused by almost any fungus.
–are commonly classified as superficial, cutaneous, subcutaneous, and systemic infections; the systemic infections are subdivided into those caused by the pathogenic fungi and those caused by opportunistic fungi.

A. Classification of Fungal Infections Commonly Found in the U.S.

Type	Disease	Causative Organism
Superficial mycoses	Tinea nigra	*Exophiala werneckii*
	Pityriasis versicolor	*Malassezia furfur*
	Piedra	*Trichosporon beigelii* (white)
		Piedraia hortae (black)
Cutaneous mycoses	Dermatophytoses	Dermatophytes (*Microsporum, Trichophyton Epidermophyton*)
	Candidiases	*Candida albicans* and related species
Subcutaneous mycoses	Sporotrichosis	*Sporothrix schenckii*
	Chromyblastocosis	*Fonsecaea, Phialophora, Cladosporium*
	Mycotic mycetoma	*Pseudallescheria boydii, Madurella*, etc.
Systemic mycoses	Pathogenic fungus infections	
	Histoplasmosis	*Histoplasma capsulatum*
	Blastomycosis	*Blastomyces dermatitidis*
	Paracoccidioidomycosis	*Paracoccidioides brasiliensis*
	Opportunistic fungus infections	
	Cryptococcosis	*Cryptococcus neoformans*
	Malessezia fungemia	*Malessezia furfur*
	Aspergillosis	*Aspergillus fumigatus*, etc.
	Zygomycosis (phycomycosis)	*Mucor, Absidia, Rhizopus, Rhizomucor*
	Candidiasis, systemic & local	*Candida albicans, Candida sp.*
	Pseudallescheriasis	*Pseudallescheria boydii*

MODIFIED FROM: Rippon, Medical Mycology

B. Diagnosis

–is based on (1) clinical suspicion; (2) direct microscopic examination of skin, mucous membrane lesions, or body fluid sediments or filtrates; (3) special stains of histologic sections; (4) cultures of body fluids, biopsy material, or specimens.

C. Clinical Suspicion of Fungal Infection Occurs When

–a patient has a flu-type infection that has lasted longer or is more severe than a viral flu.

–a patient has a chronic respiratory problem with weight loss and night sweats.

–a compromised patient with fever of unknown origin does not respond to antibacterial agents or initially responds and then gets worse. (Mixed infections occur commonly in severely compromised patients.)

–a patient with any infection with negative bacterial cultures does not respond to antibiotics.

–a patient has signs of meningitis.

–a patient has a known exposure to bird or bat guano (e.g., a spelunker) or desert sand and develops a respiratory infection.

D. Rapid Microscopic Methods for Demonstration of Fungi in Patient Specimens

–include potassium hydroxide (KOH) mount of skin scrapings, hairs, nail clippings, lesion exudates, or sputum. (KOH breaks down the human cells, although the fungus is unaffected.)

–include nigrosin or india ink mount of CSF sediment to demonstrate the encapsulated yeast *Cryptococcus neoformans.*

–include a Giemsa or Wright's stain of thick blood or bone marrow smear to detect *Histoplasma capsulatum.*

–include Calcofluor white stain with Evans Blue as a counter stain for exudates and small skin scales, which is then visualized on a fluorescent microscope.

E. Special Fungal Stains Done in the Histology Lab on Tissue Specimens

–are necessary because fungi are not distinguished by color with an H&E stain.

–include the Gomori Methenamine Silver Stain (fungi are stained dark grey to black) and the Periodic Acid Schiff Reaction (fungi stain hot pink to red).

–may also include the Gridleys stain (rose to purple fungi with bright yellow background) or the Calcofluor white stain.

F. Cultures for Fungi

–employ special fungal media depending on the specimen.

–are usually done on a single medium with antibiotics to inhibit the growth of the skin bacteria when the main concern is dermatophytes.

–are usually done on Sabouraud's medium where yeasts are expected; this is a traditional agar that encourages the growth of fungi and discourages bacterial growth.

–on specimens from normally sterile tissues are usually done on three media:

1. a basic medium (i.e., brain-heart infusion or Sabouraud's dextrose and brain heart infusion agar) with antibiotics (chloramphenicol and gentamicin).
2. an enriched medium (blood agar) with antibiotics.
3. a medium with antibiotics and a drug cycloheximide, which will inhibit most of the airborne fungal contaminants.

–usually require special processing of blood specimens since the number of organisms per milliliter is small.

G. Identification of Yeast Isolates

–involves morphologic characteristics (presence of capsule, formation of germ tubes in serum, morphology on cornmeal agar), and biochemical tests (urease, nitrate reduction and carbohydrate assimilations and fermentations).

H. Identification of Filamentous Isolates

–is based primarily on morphology.

–may also be done using an immunologic method termed exoantigen testing in which antigens extracted from the culture to be identified are immunodiffused with antiserum against known fungi.

I. Serology

–is used to identify specific antibodies to the fungi.

–is complicated by some cross-reactivity to the pathogenic fungi.

–may also use known antibodies to identify circulating fungal antigens in a patient's serum.

J. Antifungal Drugs Used for Skin, Vaginal, and Eye Infections Include

–**Nystatin**, a polyene drug not absorbed from the GI tract that is used topically, intravaginally, or orally to treat yeast infections or reduce yeast growth in the GI tract of compromised patients.

–**Griseofulvin**, an inhibitor of microtubules that is administered orally and localizes in the stratum corneum and is effective against dermatophytes but may worsen yeast infections.

–**Ketoconazole**, an imidazole used orally for yeast and dermatophyte infections; it is particularly useful in griseofulvin-resistant cases.

–**Clotrimazole and Econazole**, imidazoles used topically for dermatophytic or yeast infections.

–**Miconazole**, an imidazole used topically for dermatophytic or yeast infections.

–**Tolnaftate**, an antifungal used for dermatophytic skin (not for nail or hair) infections and that is not effective against *Candida*.

–**Haloprogin**, a topical for dermatophytes and yeasts.

–**Pimaricin**, a topical used primarily in fungal keratitis.

–**KI** (potassium iodide), given orally in milk in the treatment of subcutaneous sporotrichosis.

K. Antifungal Drugs Used in Systemic Infection Include

–**Amphotericin B (AMB)**, a polyene antifungal administered IV and that, although it does cause some nephrotoxicity, is still the drug of choice in life-threatening fungal infections.

–**5-Fluorocytosine (5FC)**, an antimetabolite that is administered orally. There are problems with the development of drug resistance. As a result, it is used primarily in combination with AMB in the treatment of *Cryptococcus* meningitis or used alone at high strength to irrigate the bladder in yeast infections.

–**Ketoconazole**, an orally administered imidazole useful in non-life-threatening systemic fungal infections. It is particularly useful in chronic mucocutaneous candidiasis. (It is ineffective against *Aspergillus*.)

–**Miconazole**, another imidazole requiring IV administration and exhibiting greater toxicity than ketoconazole.

Types of Mycoses
Superficial Mycoses

I. General Characteristics

–affect outermost layer of skin and hair.

–generally cause no cellular response to the infection.

–have primarily cosmetic symptoms.

–include: pityriasis versicolor, tinea nigra, and white or black piedra.

II. Pityriasis Versicolor

–is a fungal infection of the stratum corneum that manifests as hypo- or hyperpigmented skin patches, usually on the trunk of the body. (Color of patches varies with pigmentation of skin, exposure to sun, and severity of disease.)

–is diagnosed by KOH mount of skin scales. Wood's lamp fluorescence will be yellow.

–is treated most easily with selenium sulfide. Can be stripped off with adhesive tape or treated with a keratolyic agent.

–recurs.

–is caused by *Malassezia furfur,* which is found in skin scales as short curved septate hyphae and yeast-like cells (spaghetti and meatballs).

III. Tinea Nigra

–is a superficial infection of the stratum corneum on the palmar surfaces caused by *Exophiala werneckii,* a dematiaceous fungus.

–is characterized by black to dark-brown nonscaly, nonraised palmar lesions, misdiagnosed sometimes as malignant melanoma.

–is diagnosed by the finding of dematiaceous hyphae in characteristic lesions.

–is treated with keratolytic agents.

IV. Piedra

A. The Disease Piedra

–is a fungal infection of the hair shaft, producing hair breakage.

–is treated by cutting or shaving the hair and with topical fungicides such as bichloride of mercury.

B. Causative Agents

–has two different forms: white piedra and black piedra.

1. *Trichosporon cutaneum*

–causes white piedra.

–produces a soft, easily detached mass of hyphae, arthroconidia and blastoconidia.

–must be distinguished from nits and lice.

2. *Piedraia hortae*

–causes a black piedra.

–produces a hard, black, firmly adhered, gritty nodule that has asci and ascospores in it.

–must be distinguished from pediculosis.

Cutaneous Infections

I. General Characteristics

–may involve the skin, hair, or nails.

–may be caused by any of the dermatophytes, a homogeneous group of filamentous fungi with three genera: *Epidermophyton, Microsporum,* and *Trichophyton* or may be caused by some of the yeasts, primarily *Candida.*

–may give rise to a hypersensitive state called the dermatophytid or ID reaction as a result of circulating fungal antigens.

–are classified by the area of the body involved.

–may fluoresce under a Wood's light if caused by *Microsporum.*

–may be acquired from animals (zoophilic), in which case lesions are quite inflammatory.

–may involve the hair follicles, mandating treatment with an oral antifungal.

–may be acquired from humans (anthropophilic), in which case there is less inflammation.

–are diagnosed primarily by microscopic examination of skin, hair, or nail material mounted in 10% KOH for examination.

II. Tinea Capitis

A. The Disease, Tinea Capitis

–is also called ringworm of the scalp.

–is an infection of the skin and hair of the head.

–has pediatric and adult forms.

–has an epidemic form occurring in school children and nonepidemic forms usually transmitted by animals (zoophilic tinea capitis).

–is characterized by scaly erythematous areas of the scalp or sometimes by highly inflamed, boggy areas of the scalp called kerion.

–is diagnosed by using a Wood's light (most species of *Microsporum* fluoresce) and KOH mount of skin and plucked hairs.

B. The Forms of Tinea Capitis

1. Anthropophilic or Epidemic Tinea Capitis in Children

–is caused by *Microsporum audouinii*.

–is usually noninflammatory and produces grey patches of hair.

–is contagious through head bands, hats.

–may heal spontaneously, especially at puberty.

–is treated with griseofulvin po and a topical fungistatic agent such as boric acid to reduce infectivity. (Oral ketoconazole will work if the patient cannot take griseofulvin.) The head should be scrubbed daily to remove infectious debris. Tx may last months.

2. Zoophilic Tinea Capitis in Children

–is most commonly caused by *Microsporum canis* or *Trichophyton mentagrophytes*.

–is more inflammatory, with kerion in *T. mentagrophytes* infections.

–is usually transmitted by pets, and occasionally by farm animals, which also need to be treated.

–may have temporary alopecia, kerion, keloid, and inflammation.

–may heal spontaneously but is usually treated with oral antifungals since it is often tender to touch.

3. Black-Dot Tinea Capitis of Adults

–is usually caused by *Trichophyton tonsurans*.

–is characterized by hair breakage, followed by filling of follicles with dark conidia.

–is often a chronic infection.

–is usually treated with oral griseofulvin or oral ketoconazole.

4. Favus or Tinea Favosa

–is a severe form of tinea capitis with scutula formation and permanent hair loss due to scarring.

–is caused by *Trichophyton schoenleinii*.

–is treated with griseofulvin and removal of debris. Family members should be treated concurrently.

III. Tinea Barbae

–is an acute or chronic folliculitis of the beard, neck, or face.

–has pustular or dry scaly lesions.

–may be superinfected with bacteria.

–is treated after any bacterial infection in the same area has healed.

–is diagnosed (as are all the tineas) by KOH mount.

–must be treated with an oral antifungal like griseofulvin because the hair follicle is infected.

–is most commonly caused by:

Trichophyton verrucosum (most common causative agent in dairy areas, acquired from animals)
Trichophyton mentagrophytes
Trichophyton rubrum

IV. Tinea Corporis

–is a fungal infection of the glabrous skin. (Where the infection is limited to inside skin folds, it is usually a yeast infection rather than a dermatophytic one.)
–is characterized by annular lesions with an active border that may be pustular or vesicular.
–is treated with topicals, except in recalcitrant infections (usually *T. rubrum*).
–is most commonly caused by: *Trichophyton rubrum, Trichophyton mentagrophytes,* or *Microsporum canis.*

V. Tinea Cruris

–is an acute or chronic fungal infection of the groin, perianal and anal regions.
–is diagnosed by clinical picture and KOH mount of skin scrapings. Dermatophytic infections will show true hyphae and arthroconidia in a KOH mount.
–simulates a yeast infection. Yeast infections will show pseudohyphae and blastoconidia in a KOH mount. Yeast infections are treated with nystatin, haloprogin, topical miconazole or ketoconazole.
–is often accompanied by athlete's foot, which also must be treated.
–is treated with topical antifungals like tolnaftate, econazole, miconazole.
–is commonly caused by:

Epidermophyton floccosum
Trichophyton rubrum
Trichophyton mentagrophytes
Candida species or other yeasts

VI. Tinea Pedis

–is an acute to chronic fungal infection of the feet.
–may be superinfected with bacteria that require antibiotic treatment first.
–is most commonly caused by:

Trichophyton rubrum
Trichophyton mentagrophytes
Epidermophyton floccossum

A. Forms of Tinea Pedis

Chronic Intertriginous Tinea Pedis

–has white macerated tissue between toes; the most common form.
–is treated with imidazoles of tolnaftate, keeping feet dry (aluminum chloride) and aerated. If infection persists, use griseofulvin or oral ketoconazole.

Chronic Dry, Scaly Tinea Pedis

–hyperkeratoic scales on heel, sole, and sides of feet.
–treat with keratolytic agent like Whitfield's ointment and griseofulvin.

Vesicular Form

–treat gently to avoid massive ID reaction. Permanganate or Burow's solution to open vesicles and release antigens to surface. Griseofulvin is treatment of choice.

VII. Tinea Manuum

–is a dermatophytic infection of the hand.
–is caused by the same organisms causing tinea corporis.

VIII. Tinea Unguium

–dermatophytic infection of the nails.
–rarely is associated with paronychomycoses (swelling around the nails).
–is difficult to treat; use 2% miconazole in alcohol applied topically or long-term oral griseofulvin (over 1 year for fingernail, 15 months for toenails).

VIII. Onychomycosis

–is a nondermatophytic fungal infection of the nails.
–is most commonly caused by *Candida,* resulting in swelling and pain (paronychomycosis) in the tissue around the nail.
–is usually resistant to griseofulvin.

IX. Dermatophytid Reaction

–is an allergic reaction to circulating fungal antigens.
–is more common with severe vesicular infections.
–usually consists of a follicular reaction (papules or vesicles) at a site distant to a dermatophytid infection.
–may be exacerbated by griseofulvin treatment.

Subcutaneous Mycoses

I. General Characteristics

–are mycoses that generally start with the traumatic implantation of normally saprobic fungi and remain localized in the cutaneous and subcutaneous tissues. In some cases there is limited slow spread via the lymphatics.
–are uncommon in the U.S.; the following have been reported: sporotrichosis, mycetoma (eumycotic and actinomycotic), and chromomycosis.

II. Sporotrichosis

A. Sporothrix Schenckii

–is the causative agent of sporotrichosis.
–is a dimorphic fungus that grows at 37°C as a cigar-shaped budding yeast and at 25°C as sporulating hyphae.
–is found in/on plant materials (rose thorns, sphagnum moss, mine timbers).

B. Forms of Sporotrichosis

1. Lymphocutaneous Sporotrichosis

–is a subcutaneous, nodular, fungal disease that spreads via the lymphatics.
–is generally not painful.
–most commonly appears on limbs or extremities such as the nose.
–classically presents with a chain of lesions (lower ones often ulcerating following the lymph nodes along a limb).

–is nicknamed the rose gardener's disease in the U.S.

–is diagnosed by culture; histology is generally negative.

–is treated by potassium iodide (KI) drops in a milk menstruum.

2. Fixed Cutaneous Sporotrichosis

–differs in that the lesions do not spread via the lymphatics.

–may produce satellite lesions adjoining the initial lesion.

–occurs primarily in endemic regions.

–lesions may heal and reappear.

–is diagnosed by culture.

–is treated by KI drops in a milk menstruum.

3. Mucocutaneous Sporotrichosis

–has lesions in the nose, pharynx, and mouth that may be painful.

–is diagnosed by culture.

–is treated by KI drops in milk.

4. Pulmonary Sporotrichosis

–is not a subcutaneous infection but an opportunistic pulmonary infection.

–is primarily a problem in poverty situations and in alcoholics; it is not uncommon in large urban hospitals.

–has a chronic form resembling chronic TB in onset, chronicity, cavitation, and localization, which may be fatal.

–has an acute form with major involvement of the lymph nodes rather than the lung parenchyma. This form develops rapidly but may resolve spontaneously.

–is diagnosed by both culture and serology (generally immunodiffusion).

–is treated by AMB or surgical excision of major cavities.

III. Eumycotic Mycetoma

A. Causative Agents

Pseudallescheria boydii, Madurella

–all are filamentous fungi.

–all live in soils and on vegetation.

–enters by wound; usually occurs in rural, third world, agricultural workers in the tropics.

B. The Disease Eumycotic Mycetoma

–is caused by true fungi.

–is a subcutaneous disease characterized by swelling, drainage through sinus tracts and granules.

–is usually characterized by a flat (but swollen) surface around the sinus tracts.

–spreads contiguously to adjoining region and will invade deep tissues and bone.

–rarely disseminates, except for shoulder and back lesions.

–is diagnosed by microscopic exam of the granules and sometimes by culture of the granules. Filaments 4 to 10 microns wide with clubbing at the periphery of the granule are characteristic.

–is diagnosed in the tropics with immunodiffusion and counterimmunoelectrophoresis.

–is difficult to treat. Surgical draining and debridement of diseased tissue are used along with antifungals. AMB is not effective. Ketoconazole and miconazole have been effective in some cases.

IV. Chromoblastomycoses

–is one of a group of infections called chrommomycoses, which are grouped together because they are all caused by dematiaceous fungi.

–is the most common form of chrommomycoses.

–occurs most commonly outside the U.S. in poor, male rural workers where nutrition is poor and clothing scant.

A. The Causative Agents of Chromoblastomycoses

–are most commonly caused by one of three genera of dematiaceous fungi:

Phialophora
Fonsecaea
Cladosporium

–their taxonomy is still unresolved.

–are common soil organisms. *Cladosporium* normally is the number one air contaminant.

B. Forms of Chromoblastomycoses

1. Chromoblastomycosis (Verrucous Dermatitis)

–is the most common form of chromomycosis.

–is a localized subcutaneous infection of the limb or shoulder.

–begins with traumatic implantation of the spores.

–may be prevented by good nutrition.

–has colored lesions that start out scaly and become raised cauliflower-like lesions.

–is generally not painful except when bacterial superinfection occurs.

–is diagnosed by finding dark sclerotic bodies in the infected tissue.

–is treated in early stages by surgery, electrodessication or cryosurgery and with AMB directly into lesions. Oral ketoconazole may be effective.

2. Chromomycosis of the Brain

–is a fatal and rare infection.

–is diagnosed at postmortem and rarely treated.

3. Phaeomycotic Cyst

–is a cutaneous, subcutaneous, or intramuscular cyst.

–is often hard.

–may ulcerate and exude pus; it is a brown pigmented fungi and has sclerotic bodies.

–is treated with surgery and 5-Fluorocytosine if cyst has ruptured; heat may help.

–may spread if patient becomes debilitated or is treated with steroids.

Systemic Mycoses: The Pathogens

I. General Characteristics

–are caused by dimorphic fungi.

–are also sometimes known as the "deep mycoses."

–are mycoses that affect internal organs and may disseminate to multiple sites of the body.

–can be further subdivided into those infections caused by:

1. pathogenic fungi
2. opportunistic fungi

A. Pathogenic fungi causing systemic disease include:

Blastomyces
Histoplasma
Coccidioides
Paracoccidioides-S America

B. Opportunistic fungi causing systemic disease include:

Candida
Malassezia
Cryptococcus
Geotrichum
Aspergillus
Mucor, Rhizopus, Absidia, Rhizomucor
Pseudallescheria
Sporothrix

C. Differences between pathogenic fungi and opportunists:

–opportunists have low virulence, and pathogens can invade and cause disease in a
   healthy host or in a compromised host.
–opportunists will infect only compromised patients.

D. The dimorphic fungal pathogens in the U.S. (*Histoplasma, Coccidioides,* and *Blastomyces*):

–are found in the human body as the yeast form (*Histoplasma* and *Blastomyces*) or
   spherule form (*Coccidioides*).
–grow in lab cultures at 30°C or lower as filamentous forms.
–are found in sand, soil, decaying organic material, or bird or bat feces.
–are endemic to specific geographic regions. (Medical history is important.)
–all produce airborne spores.

E. Forms of the diseases are somewhat similar for all three diseases:

a. Acute, self-limited pulmonary (asymptomatic to quite severe) involvement
b. Chronic (pulmonary or disseminated)
c. Disseminated

F. The patients are:

–primarily people who are exposed to contaminated sand or soil (farmers, bulldozer
   operators, archeologists, golfers in the southwestern U.S.) or

–people exposed to starling, chicken, or bat feces (spelunkers, cleaners of chicken coops, building demolition teams).

II. Histoplasmosis

–are granulomatous fungal infections caused by *Histoplasma capsulatum*; 95% are inapparent, subclinical, or self-resolving. The disease leaves residual calcification in the lungs.

–is one of the most common diseases in the U.S. (Acute infections, known as the fungus flu usually self-resolve and are usually undiagnosed.)

–must be differentiated from flu, pneumonia.

A. *Histoplasma capsulatum*

–is a dimorphic fungus that is a facultative intracellular parasite localizing in monocytic cells circulating throughout the reticuloendothelial system (RES).

–is found in soil enriched with bat or bird (particularly chicken or starling) guano.

–can be isolated from most areas around the world; the major endemic region lies in the drainage areas of the Ohio, Missouri, and Mississippi rivers and the St. Lawrence Sea Way. One hundred percent of the chicken coops in some areas of Missouri are positive for *H. capsulatum*.

B. The forms of histoplasmosis

–can be classified into acute, chronic, or fulminant.

–also classified as localized or disseminated.

–depend largely on the health, lung structure, and immune system of the host and the dose of the inoculum.

1. Acute histoplasmosis

–ranges from asymptomatic to a severe but self-resolving disease.

–symptoms are flu with fever, aches and pains along with a nonproductive cough, pleuritic pain, shortness of breath, and hoarseness. In more severe disease, in addition to the above, there are fever, night sweats, weight loss, some cyanosis, and occasional hemoptysis.

–has an initial incubation period, usually 2 to 23 days; and most acute infections last about 2 weeks.

–shows a highly variable x-ray pattern; usually includes multiple lesions scattered throughout the lung accompanied by hilar lymphadenopathy. Lungs may show buck shot, popcorn lesions. Calcification extremely common on healing, especially in the young. Characteristically both lungs are involved.

–is almost always a pulmonary disease; however a transient, hematogenous spread of the fungal cells occurs via the macrophages.

–is treated in healthy Caucasian, nonpregnant individuals with bed rest and good nutrition.

2. Chronic pulmonary histoplasmosis

–may change rapidly from an acute to a chronic infection, or the acute infection may heal and reactivate as a chronic disease.

–symptoms are long term with no signs of improvement.
–usually produces a variety of x-ray patterns, including cavitation (high incidence of dissemination), histoplasmoma in a hypersensitive individual, or coin lesions.
–must be treated with antifungals such as ketoconazole in non-life-threatening chronic infections.

3. Chronic disseminated histoplasmosis

–will show numerous organisms in macrophages.
–may not have apparent pulmonary symptoms.
–generally patients present with low-grade fever, weight loss, weakness, and focal destruction of organs.
–will often have oral/pharyngeal lesions of mucous membrane lesions.
–treated with AMB (life threatening) or KTZ (maintenance).

4. Disseminated histoplasmosis (fulminant cases)

–occurs in individuals with underlying immune cell defects (e.g., AIDS patients or any patient with T-cell deficits, lymphoma patients) and in children less than 1 year who appear to have an RES defect (fulminant disease of childhood).
–causes Addison's disease in about 50% of the cases.
–symptoms are:
    mucous membrane lesions or focus of infection other than lungs;
    prominent hepatosplenomegaly;
    decrease in WBC, Hgb, platelets, DIC;
    pulmonary symptoms (although these may be lacking);
    anemia, leukopenia;
    weight loss;
    must be treated aggressively, usually with AMB; poor prognosis.

C. Diagnosis of histoplasmosis (multiple infections may exist, especially in AIDS patients) is based on:

–microscopic examination of thick blood, bone marrow smear, buffy coat, or liver biopsy stained with Wright's or Giemsa stain;
–microscopic exam of sputum;
–culture of sputum or bronchial washings for dimorphic fungus; and
–culture of blood for dimorphic fungi.

1. Cultures of *Histoplasma capsulatum*

–at 25°C produce filamentous, white-to-brown colonies.
–at 25°C produce hyphae with small, tear-shaped microconidia and large round tuberculate macroconidia.
–at 37°C produce creamy white yeast colonies of nondescript small cells with a narrow neck between buds and mother cells.

2. Definitive identification of cultures is based on:

–conversion from yeast form to the filamentous form with microscopic demonstration of typical sporulation.

–an alternative method called exoantigen testing.

–using soluble antigens from the fungus in an immunodiffusion.

3. Diagnostic serology includes

–complement fixation
–immunodiffusion
–radioimmunoassay

D. Skin testing for histoplasmosis

–cannot be used to diagnose active disease.

–can be used to demonstrate previous exposure to the antigen.

–a negative test is a poor prognostic sign in a patient with known active histoplasmosis.

–if done before serology can boost antibody levels.

III. Blastomycosis (North American)

–includes pulmonary, disseminated, or cutaneous fungal diseases caused by *Blastomyces dermatitidis*.

–lesions (suppurative and granulomatous) may be found most frequently in the lungs with most common dissemination site the skin, followed by bones and other organs.

–may also cause adrenal insufficiency in disseminated disease.

A. *Blastomyces dermatitidis*

–is dimorphic.

–is found in the tissues as large, mainly free, yeast with a double refractile wall and broad-based buds.

–is endemic mainly to the eastern U.S.

–is found as a mold in soil along water ways and in decaying wood.

–most likely is inhaled as conidia and transformed into the yeast form in the lung.

B. The forms of blastomycosis

–can be classified into acute or chronic and localized or disseminated.

–are more likely to need treatment than acute histoplasmosis.

–are dependent on the underlying state of health of the patient.

1. Acute pulmonary blastomycosis

–may resolve without treatment; however KTZ is being used in uncomplicated, non-life-threatening cases.

2. Chronic pulmonary blastomycosis

–usually has suppurative or granulomatous lesions in the upper lobe.

–is often misdiagnosed as carcinoma.

–rarely calcifies.
–most often presents with infiltrative pattern.
–generally no cavitation.
–needs treatment.

3. Dissemination of primary pulmonary blastomycosis

–generally occurs in individuals with low stimulation index to the antigen and negative skin test.
–organisms are carried by the macrophages to remote sites.
–defined by lesions found in the lung and the skin, which is the most common site of dissemination.
–a microscopic KOH wet mount of material from the pustular edges of these skin lesions is the fastest way to establish a tentative diagnosis.
–is treated, when life threatening, with AMB.

C. Differential diagnosis of blastomycosis is based on:

1. Radiology: infiltrative pattern, nodular pattern, and single lesion resembling neoplasm are most common;
2. Direct microscopic exam—pus or skin scrapings or sputum examined after KOH digestion;
3. Culture for fungi;
4. Pathology: *Blastomyces* may be extracellular or found in giant cells; suppurative reaction is most common;
5. Serology, which includes:
   –complement fixation and immunodiffusion, although cross-reactions are still a problem.
   –radioimmunoassay, which is more sensitive.

D. Skin test: for demonstration of anergy.

IV. Coccidioidomycosis

–includes primary, acute, self-limited fungal infections to chronic and disseminated diseases.
–occurs in the lower Sonoran desert of southern California, Arizona, New Mexico, and western Texas.
–is commonly known as valley fever.
–is caused by the dimorphic fungus *Coccidioides immitis*.

A. The organism *Coccidioides immitis*

–is found as arthroconidia in the sand.
–is found in the human body as large spherules within which endospores develop. Endospores are released when mature and enlarge to form new spherules.
–causes major epidemics after dust storms.

B. The forms of coccidioidomycosis

–range from asymptomatic to acute to chronic forms and may remain localized to the lungs or may disseminate.

–depend on the underlying health and immune system of the patient.

–include:

1. Asymptomatic.
2. Primary self-limited respiratory infection (duration few days to several weeks). The symptoms are:

   fever;

   cough;

   chest pain;

   body aches and pain;

   anorexia;

   joint pain.

   –erythema nodosum or multiforme, an allergic RX, which if present is a good prognostic sign.

   –lung changes: pneumonic infiltrate to large cavitary lesions.

   –generally is treated with bed rest, good nutrition. Treat with antifungal reagent if pregnant, diabetic, immunosuppressed or black.

   –may require surgical removal of large lesions in benign residual disease to prevent later reactivation.
3. Disseminated disease (1% to 10%)

   –involves many tissues and organs including brain, bones, and skin.

   –is evidenced by:

   persistent elevated sedimentation rate;

   persistent elevated complement-fixation titer;

   persisent and increasing fever and malaise.

   –is treated with AMB for life-threatening disease or maintenance on KTZ for continuously immunocompromised patients.

C. Diagnosis of coccidioidomycosis is based on:

1. Direct exam: scrapings of any lesions, sputum, or bronchial washings.

   –special fungal stains on tissues (showing: spherules with endospores; no budding yeasts)
2. Culture: sputum, bronchial washings, biopsy, scrapings; this is a highly infectious agent.
3. Diagnostic serology including

   1. a tube precipitin test that measures IgM.
   2. a complement-fixation test that measures IgG.
   3. latex particle agglutination and immunodiffusion tests are used as a screening tool in endemic areas and can detect 93% of the cases.

D. Skin test that becomes positive early in infection.

Systemic Mycoses: Opportunistic Mycoses

I. Infections

–are caused by endogenous or ubiquitous organisms of low inherent virulence that cause infection in debilitated patients.

–range from annoying or painful to rapidly fatal systemic infections.

–are increasing as the number of compromised patients increase.

–are rarely serious in well-nourished, drug-free, healthy people.

–may be caused by any fungus if a patient is immunocompromised (e.g., the common edible commercial mushroom has caused fatal infection in at least one severely neutropenic leukemic child).

–know:

1. what opportunistic infections a specific patient is at high risk for and their signs and symptoms.
2. that it is important to inform the laboratory that your patient is compromised.
3. that it is important to inform the nursing staff to be aware of the risks and monitor the patient for early signs of infection.

–are caused most commonly by *Candida, Cryptococcus, Geotrichum, Aspergillus,* and *Rhizopus, Mucor,* and *Absidia (Zygomycota).*

II. Symptoms and Predisposing Conditions

Basic Presenting Symptoms	Underlying Conditions	Diseases You Should Think About
vaginitis	antibiotic usage, pregnancy	*Candida* vaginitis
facial swelling	diabetes	rhinocerebral mucormycosis
fever w/o pulmonary symptoms	IV catheter, lipid supplement	fungemia (*Candida, Malassezia*)
fever, pain on urination	urinary catheter	urinary Candidiasis
difficulty swallowing	AIDS	esophageal Candidiasis
meningeal symptoms	AIDS, diabetic, IC pt*, Hodgkins, any pt	Cryptococcal meningitis, If exposure is right-Cocci meningitis, or disseminated invasive Aspergillosis, Candidal cerebritis
pulmonary symptoms	IC pt particularly if neutropenic	invasive Aspergillosis
	AIDS	Histoplasmosis, Coccidioidomycosis
	alcoholics, urban	Sporotrichosis, pulmonary
hemoptysis	Pt with previous lung damage, part. cavities	Aspergilloma
endocarditis	Drug addicts	*Candida* endocarditis
enteritis (often with anal puritis)	pt on antibiotics	*Candida* enteritis (irritable bowel syndrome)
whitish covering in mouth	premies, kids on antibiotics	*Candida* thrush
corners of mouth sore	elderly with malnourishment	perleche
gums sore	dentures	denture stomatitis or allergy to antifungal used to treat above
skin lesions, endoophthalmitis	pt with IV catheter	Candidemia

*IC pt = immunocompromised patients

III. Candidiases (Candidoses, or Old Term: Moniliases)

–are acute to chronic fungal infections involving mouth, vagina, skin, nails, bronchi or lung, alimentary tract, bloodstream, urinary tract, and less commonly the heart or meninges.

–are caused by *Candida albicans* or other species of *Candida.*

–are predisposed by extremes of age, wasting disease, nutritional diseases, excessive moisture, pregnancy, diabetes, long-term antibiotic and steroid usage, and indwelling catheters.

A. *Candida albicans*

–is the most common of the numerous species of *Candida*.
–is part of the normal flora of the skin, mucous membranes, and GI tract along with other species of *Candida*.
–forms elongated "budding forms" called pseudohyphae; these are often seen in clinical material along with true hyphae and blastoconidia and yeast cells.
–is identified in the laboratory by:
> formation of germ tubes when yeast isolates are incubated in serum;
> formation of hyphae and chlamydoconidia at 25°C on cornmeal-Tween 80 agar;
> formation of blastoconidia at septae of pseudohyphae at 30°C;
> carbohydrate assimilations if necessary.

Speciation of other strains is done morphologically and chemically.

B. Common forms of Candidiases

1. Oral thrush
   –is a yeast infection of the oral mucocutaneous membranes.
   –manifests itself as white curd-like patches in the oral cavity.
   –occurs in premies, older infants on antibiotics, immunosuppressed patients on long-term antibiotics, AIDS patients.
   –usually requires no diagnostic work up; if in doubt, do a KOH mount and examine microscopically.
   –is treated with oral nystatin, miconazole, or ketoconazole.
2. Vulvovaginitis or vaginal thrush
   –is a yeast infection of the vagina.
   –manifests itself with a thick yellow-white discharge, burning sensation, curd-like patches on the vagina mucosa, and inflammation of the peritoneum.
   –is predisposed by diabetes, antibiotics, oral contraceptives, and pregnancy.
   –has a tendency to recur.
   –is diagnosed with a KOH mount.
   –is treated with intravaginal nystatin or imidazole drugs (occasionally orally or topically).
3. Cutaneous candidiasis
   –involves nails, skin folds, or groin region.
   –may be eczematoid or vesicular and pustular.
   –is predisposed by moisture.
   –is diagnosed by KOH mount.
   –is treated with weight reduction, adequate aeration, and topical nystatin or imid-azoles.
4. Alimentary tract disease, including esophagitus
   –is usually an extension of oral thrush.
   –is found in AIDS patients and other immunosuppressed patients, particularly those on long-term antibiotics.

–is reduced in highly susceptible populations by fungal prophylaxis.

–is treated with AMB or oral nystatin, oral ketoconazole or clotrimazole.

5. Systemic candidiases may occur almost anywhere in the body:

–candidemias or blood-borne infections occur most commonly in patients with indwelling catheters (and may, in immuno*competent* patients be resolved spontaneously by removing the catheter).

–endocarditis occurs in patients who have manipulated or damaged valves or in intravenous drug abusers.

–bronchopulmonary disease (usually manifested by persistent cough) occurs in patients with chronic lung disease.

–cerebromeningeal disease may occur in compromised patients.

6. Systemic candidiasis

–is diagnosed by demonstration of the presence of pseudohyphae or true hyphae in the tissues; or by the culture of the organisms from normally sterile areas of the body such as the blood or brain; or by serologic methods demonstrating high levels of *Candida* precipitins or antigens; or chromatographic methods.

–may be manifested by candiduria, maculonodular skin lesions, or endophthalmitis.

–is treated with amphotericin B in most cases.

7. Chronic mucocutaneous candidiasis

–is a chronic, often disfiguring, infection of the epithelial surfaces of the body.

–is diagnosed microscopically and by the lack of cell-mediated immunity to *Candida* antigens.

–usually occurs in patients with T-cell deficiencies and in some cases also endocrinopathies.

–is treated with oral ketoconazole.

C. Diagnostic methods vary depending on the site of the infection.

–Cultures of areas of the body normally colonized by *Candida* are not useful diagnostic tool, whereas cultures of normally sterile parts of the body are significant.

–Skin testing with *Candida* antigen is useful as a general test for T-cell functioning in children since exposure to *Candida* is universal.

IV. *Malassezia Furfur* Septicemia

–is a blood-borne infection caused by the lipophilic skin organism *Malassezia furfur*.

–occurs in patients (primarily neonates) on intravenous lipid emulsions.

–is diagnosed by culturing blood on fungal media that is either lightly overlayered with sterile olive oil or has lipids incorporated into the media.

–may resolve by halting the lipid supplements.

V. Cryptococcosis

–includes subacute or chronic fungal infections involving lungs, meninges, or, less commonly, skin, bones, and other tissues.

–presents clinically most commonly as meningitis.

A. *Cryptococcus Neoformans*

–is a yeast possessing an antigenic polysaccharide capsule.

–may be isolated from fruit, milk, vegetation, and soil.

–is also associated with pigeon excretia and is an occupational hazard to pigeon handlers.

–is considered to be an opportunist in the presence of underlying disease in patients with Hodgkin's, leukemias, or leukocyte enzyme deficiency diseases.

B. *Forms of Cryptococcal Disease*

1. Pulmonary Infections

–are usually asymptomatic and self-resolving.

–in their more rare and fulminant forms are highly variable but may resemble pneumococcal pneumonia.

–are diagnosed by wet mount and culture.

–may or may not require treatment, depending on the severity of the disease and the underlying condition of the patient.

2. Meningitis or Meningoencephalitis

–presents most commonly as headache of increasing severity, usually with fever followed by typical meningitis signs.

–is diagnosed by detection of the capsular material in the CSF by the cryptococcal antigen latex agglutination test.

–may be diagnosed in 50% of cases by demonstration of encapsulated yeast in CSF sediment on wet mount in nigrosin or india ink.

–is confirmed by isolation of *C. neoformans* by culture of CSF.

–is treated by AMB usually in conjunction with 5-fluorocytosine.

C. Diagnosis

–methods vary depending on the site of the infection.

VI. Aspergilloses

–are a variety of infectious and allergic diseases caused by *Aspergillus fumigatus* and a variety of other species of *Aspergillus*.

A. *Aspergillus*

–is a ubiquitous filamentous fungus whose airborne spores are constantly in the air.

–is recognized both in tissue and in cultural growth by its characteristic septate, dichotomous branching.

–produces conidial heads with numerous conidia.

–is an opportunistic organism.

–*fumigatus* and *flavus* are the most common species.

B. Forms of aspergilloses

1. Allergic bronchopulmonary aspergillosis (ABPA)

–is an allergic disease in which the organism colonizes the mucous plugs formed in the lungs but does not invade lung tissues.

–is generally a sequela to asthma and presents as chronic, severe asthma.

–is diagnosed by immunodiffusion and direct microscopy and culture of the sputum.

–is treated with glucocorticosteroids.

–is not an easily resolvable problem as *Aspergillus* often is actually growing on the surfaces of the lung.

2. Aspergilloma

–is a roughly spherical growth of *Aspergillus* in preexisting lung cavities. The growth does not invade the lung tissues.

–presents clinically as recurrent hemoptysis.

–is a sequela to cavitary lung disease like tuberculosis.

–is diagnosed by immunodiffusion and x-ray where the lesion and the common air space above it is shown to shift when the patient is x-rayed upright and supine.

–is generally removed surgically and the diagnosis confirmed microscopically.

–generally has a good prognosis if properly treated.

3. Invasive aspergillosis

–occurs most commonly during severe neutropenia in leukemia and transplant patients.

–most commonly presents as fever of unknown origin in patients with fewer than 500 neutrophils/mm^3 and pneumonia. It may start as sinusitis. From either the sinuses or the lungs it disseminates to any part of the body.

–is diagnosed by microscopy and culture of lung biopsy material.

–is treated aggressively with AMB.

–has a high fatality rate unless neutrophil numbers become elevated.

C. Diagnostic methods vary depending on whether the disease is allergic or infectious and on the site.

D. Skin testing is not used. Resolution of infections with *Aspergillus* is not T-cell dependent but appears to reside primarily with the phagocytic cell lines.

VII. Zygomycosis (Phycomycoses)

–are infections caused most commonly by the genera *Rhizopus, Absidia, Mucor,* and *Rhizomucor,* which belong to the order Mucorales.

–include rhinocerebral forms in acidotic diabetic patients, thoracic forms in leukemia and lymphoma patients, abdominal-pelvic disease in malnourished patients, and cutaneous infections in leukemia patients.

A. *Rhizopus, Absidia, Mucor,* and *Rhizomucor*

–are all aseptate.

–are all rapid growers.

–all have a predilection for invading blood vessels.

B. Rhinocerebral Zygomycosis

–is the most common form of zygomycosis.

–occurs in the acidotic diabetes patient.

–presents with facial swelling and blood-tinged exudate, fluid in turbinates, reddish eye discharge, mental lethargy, fixation of pupils, etc.

–must be diagnosed rapidly, usually by KOH mount of necrotic tissues and exudates from eye, ear, or nose.

–is a rapidly fatal infection.

–is rarely treated successfully; treatment consists of control of diabetes, surgical debridement, and aggressive AMB treatment.

Review Test

MYCOLOGY

DIRECTIONS: For each of the questions or incomplete statements below, one or more of the answers or completions given is correct. Choose answer:

    A. if only **1**, **2**, and **3** are correct
    B. if only **1** and **3** are correct
    C. if **2** and **4** are correct
    D. if only **4** is correct
    E. if all are correct.

4.1. Fungi:
1. are prokaryotic cells.
2. have a cell wall containing complex carbohydrates like chitin.
3. are auxotrophic in their metabolism.
4. have ergosterol as their major membrane sterol.

4.2. Skin lesions are not uncommon in:
1. Disseminated blastomycosis.
2. Chronic mucocutaneous candidiasis.
3. Disseminated candidiasis.
4. Piedra.

4.3. Superficial mycoses include:
1. Blastomycoses.
2. Chromomycoses.
3. Athlete's foot.
4. Pityriasis versicolor.

4.4. Zygomycota (phycomycetes):
1. have broad aseptate hyphae.
2. one major predisposing condition to infection is diabetes.
3. include *Rhizopus, Absidia, Mucor.*
4. are dimorphic.

4.5. Stains that differentiate fungi are:
1. Calcofluor white.
2. GMS (Gomori Methenamine Silver Stain).
3. PAS (Periodic Schiff Reaction).
4. H & E (hematoxylin and eosin).

DIRECTIONS: Choose *all* correct answers. (May have more than one correct.)

A. Septae
B. Hypha
C. Mycelium
D. Dimorphic fungi
E. Yeast

4.6. A mass of fungal filaments.
4.7. Cross walls.
4.8. A filamentous fungal subunit.

A. Miconazole
B. Ketoconazole
C. Griseofulvin
D. Amphotericin B
E. Nystatin

4.9. Oral antifungal that inhibits microtubule formation and is used to treat dermatophytic infections.
4.10. Polyene antifungal for serious fungal infections.
4.11. Fungistatic imidazole used orally for systemic or superficial infections.

A. *Candida*
B. *Epidermophyton*

170

C. *Trichosporon*
D. *Trichophyton*
E. *Microphyton*

4.12. Dermatophytes.

4.13. Causative agent of piedra.

4.14. Causative agent of yeast skin infections that may resemble some dermatophytic infections.

A. Sporotrichosis
B. Chromomycosis
C. Eumycotic mycetoma
D. Actinomycotic mycetoma
E. None of the above.

4.15. A subcutaneous *fungal* disease characterized by swelling, sinus tracts, and granules.

4.16. The finding of sclerotic bodies in the tissue is diagnostic.

4.17. Most commonly a cutaneous or lymphocutaneous disease treated with KI.

4.18. Causative agent is *Sporothrix schenckii*.

A. *Blastomyces dermatitidis*

B. *Sporothrix schenckii*
C. *Histoplasma capsulatum*
D. *Coccidioides immitis*
E. *Aspergillus fumigatus*

4.19. Dimorphic fungus.

4.20. Small, nondescript yeast form is found in inside monocytic cells.

4.21. Environmental form consists of hyphae that break up into arthroconidia.

4.22. Tissue form is large yeast with a thick cell wall and broad-based buds.

A. *Aspergillus*
B. *Cryptococcus*
C. *Candida*
D. *Malassezia*
E. *Sporothrix*

4.23. Septicemias caused by this organism are found primarily in neonates on intravenous lipid emulsions.

4.24. Encapsulated yeast.

4.25. Invasive disease with this organism occurs primarily under conditions of severe neutropenia.

Answers and Explanations

MYCOLOGY

4.1. C. Fungi are eutanyotic cells with chitin containing cell walls and ergosterol containing membranes that are heterotrophic.

4.2. Frequent site of disseminated *Blastomyces* and *Candida* is the skin. CMC affects all epithelial surfaces. Piedra is superficial.

4.3. D. Pityriasis is the only superficial mycoses; all others are cutaneous or deeper.

4.4. A. All aseptate fungi are classed in the Zygomycota. In diabetic patients, they may cause rhinocerebral infections called phyco-, zygo- or mucormycoses. *Rhizopus, Absidia,* and *Mucor.*

4.5. A. Fungal cell carbohydrates bind Calcofluor and fluoresce while background tissues do not. Silver stains fungi black; counter stain green or H & E PAS stains the fungi hot pink-red. H & E does not differentiate fungi.

4.6. C. Mycelium is defined as a mass of hyphae.

4.7. A. The cross walls of hyphae are called septations or septae.

4.8. B. The fungal subunit called a hypha is a filamentous structure with or without cell walls.

4.9. C. Griseofulvin.

4.10. D. Amphotericin B is still primary drug of choice for systemic infection.

4.11. B. Ketoconazole is the most common oral imidazole used for systemic non-life-threatening disease. (Miconazole is used IV.)

4.12. B and **D.**

4.13. C.

4.14. A.

4.15. C. Eumycotic indicates fungal infection, responds to antifungal drugs.

4.16. B. The finding of dematiacious yeast-like forms with sharp planar division lines is a characteristic finding of chromomycoses.

4.17. A. Sporotrichosis spreads along lymphatics but rarely disseminates in healthy, well-nourished adults.

4.18. A.

4.19. A, B, C, and **D.** All are dimorphic.

4.20. C. *Histoplasma* is an intracellular parasite circulating in RES.

4.21. D. *Coccidioides* is found in desert sand, primarily as arthroconidia and hyphae.

4.22. A. *Blastomyces* has a double refractile wall and buds with a broad base of attachment to the mother cell.

4.23. D. *Malassezia furfur* is a lipophilic fungus that is found on skin and causes septicemia in patients.

4.24. B.

4.25. A. *Aspergillus.*

172

5

A Review of Immunology

Overview

Immunity

–may be defined as an "altered state" of responsiveness to a specific substance induced by prior contact with that substance.

Categories of Immunity

–natural immunity.
–acquired immunity.

Natural Immunity

–is nonspecific.
–is present from birth.
–consists of various barriers to external insults: skin, mucous membranes, macrophages, monocytes, polymorphonuclear leukocytes (PMNs), eosinophiles, and their contents.

Acquired Immunity

–is specific.
–is expressed after exposure to a given substance.
–involves specific receptors on lymphocytes and the participation of macrophages for its expression.

Two Categories of Acquired Immunity

–humoral immunity mediated by serum components, i.e., antibodies.
–cellular immunity mediated by cells, i.e., whole, viable lymphocytes.

Immune System

–is equivalent to the lymphoid system.
–consists of the central lymphoid organs and peripheral lymphoid organs.

Central Lymphoid Organs

–are the locale for the maturation of lymphoid cells.
–consist of the bone marrow and thymus.

Peripheral Lymphoid Organs

–are the locale for the reactivity of lymphoid cells.
–include the spleen, lymph nodes, and lymphatic channels.

Cells of the Immune System

–include the white blood cells of the body.
–there are approximately 8,000 WBC/mm^3 blood composed of:

–Granulocytes (50% to 80% of WBC)
–Lymphocytes (20% to 45% of WBC)
–Monocytes, macrophages (3% to 8% of WBC)

Development of the Immune System

–involves the maturation of pluripotential stem cells in the thymus or the bone marrow
  into T cells and B cells, respectively.
–includes the generation of specific receptors on the cell surface of the B and T cells.

Pluripotential Stem Cell Sources

–embryonic yolk sac.
–fetal liver.
–adult bone marrow.

B Cells

–mature in the bursa of Fabricius or in the Bursal equivalent (bone marrow?) in
  humans.
–are involved in the generation of humoral immunity.
–have specific immunoglobulin receptors on their surface for antigen recognition.
–mature into antibody-producing plasma cells.
–are sessile and located predominantly in the germinal centers of lymph nodes and
  spleen.

T Cells

–mature in the thymus.
–are involved in "helping" B cells become antibody-producing plasma cells.
–have specific, distinct receptors on their surface for antigen recognition.
–are involved in cell-mediated immunity.
–participate in the suppression of the immune response.
–are the predominant cells in the circulation (95%).
–are found in the paracortical and interfollicular areas of the lymph node and spleen.

Physiology of Immunity

–involves a series of events that will culminate in B-cell or T-cell activation and response to the introduction of a foreign entity into the circulation.
–includes the "processing" of this entity by a macrophage or a B cell.
–this is followed by the recognition of the foreign entity by specific and preformed
  receptors on the B cell and the T cell.
–the cells are stimulated to proliferate by a series of soluble signals between the
  macrophages, B cells and T cells (interleukins).
–the cells will undergo blast transformation and a series of mitotic divisions that will
  lead to the generation of plasma cells, producing immune globulins and/or sensitized T
  lymphocytes capable of interacting with the original stimulus.

Review Test

OVERVIEW

DIRECTIONS: For each of the questions or incomplete statements below, *one* or *more* of the answers or completions given is correct. Choose answer:

A = 1, 2, and 3 correct
B = 1 and 3 correct
C = 2 and 4 correct
D = 4 correct
E = all correct

Questions 5.1 through 5.3:

1. Bone marrow
2. Spleen
3. Thymus
4. Lymph nodes

5.1. The central lymphoid organs.
5.2. The peripheral lymphoid organs.
5.3. The lymphoid system.

5.4. Cells involved in the immune system:

1. Granulocytes.
2. Lymphocytes.
3. Monocytes.
4. Macrophages.

5.5. Pluripotential stem cells arise from:

1. Fetal liver.
2. Embryonic yolk sac.

3. Adult bone marrow.
4. Adult spleen.

5.6. B cells:

1. Arise in the bursa of Fabricius or bursa equivalent.
2. Are found in germinal centers of lymph nodes and spleens.
3. Are in the plasma cell family.
4. Are involved in humoral and cell-mediated immunity.

5.7. *T cells*:

1. Arise in the bone marrow.
2. Mature in the thymus.
3. Are predominantly recirculating lymphocytes.
4. Are involved in humoral and cell-mediated immunity.

Answers and Explanations

OVERVIEW

5.1. B. Central organs are the seeding organs and maturation locales of the lymphatic (immune) system and would include the bone marrow and thymus.

5.2. C. Peripheral organs are the sites of residence of the mature lymphocytes: spleen, lymph nodes, and lymphatic channels.

5.3. E. The lymphoid system would contain both central and peripheral lymphoid tissue.

5.4. E. PMNs (granulocytes), lymphocytes, macrophages, monocytes, and all their derivatives (Kupffer's cells of liver, Langerhans' cells of skin) are involved in the generation of the immune response.

5.5. A. Pluripotential stem cells, i.e., cells giving rise to reticulocytes, monocytes, lymphocytes, etc., come from fetal liver, embryonic yolk sac, and adult bone marrow. The adult spleen is not a hematopoietic organ in humans.

5.6. A. B cells, or bursal cells, are not involved in cell-mediated immunity but do have all the other characteristics listed.

5.7. E. T cells hve a regulatory function in humoral immunity, derive from bone marrow stem cells, mature in thymus, and are the main type of lymphocyte in the circulation.

Antigens

Antigen (Immunogen)

–any entity capable of evoking a specific immune response.

Properties of Antigen (Ag)

–*immunogenicity:* the capacity to stimulate production of protective humoral or cellular immunity (i.e., production of antibody or specific T cells, respectively).

–*specific reactivity* with the antibody and/or T cell produced.

Factors Determining Antigenicity

–*nonself*—Ag must be recognized as foreign; foreign proteins are excellent antigens.

–*size*—Ag must have a minimum weight to be recognized, usually 10 kilodaltons molecular weight.

–*shape*—tertiary and quartenary structure enhances and defines extent of antigenicity.

Antigenic Determinant - Epitope

–that restricted portion of an Ag molecule that determines the specificity of the reaction with antibody, and reacts with the antibody-combining site.

–generally contains 4 to 6 amino acid or sugar residues.

Hapten

–A small foreign molecule that is not immunogenic but can bind to an antibody molecule already formed to it of and by itself.

–can be immunogenic if coupled to a sufficiently large *carrier* molecule.

[Example: 2,4 Dinitrophenol (DNP) is a hapten, too small to evoke an antibody (Ab) response, but DNP coupled to ovalbumin (OA) will evoke a very good Ab response to both entities.]

Review Test

ANTIGENS

DIRECTIONS: For each of the questions or incomplete statements below, *one* or *more* of the answers or completions given is correct. Choose answer:

A = 1, 2, and 3 correct
B = 1 and 3 correct
C = 2 and 4 correct
D = 4 correct
E = all correct

5.1. Of the following, which are factors in determining antigenicity?
1. Size of antigen.
2. Shape of antigen.
3. Tertiary structure.
4. "Foreignness."

DIRECTIONS: Each set of lettered headings below is followed by a list of numbered words or phrases. Choose answer.

A. Antigen
B. Hapten
C. Both
D. Neither

5.2. Can bind with preformed Ab.

5.3. Can elicit Ab response *per se.*
5.4. Are "foreign" relative to host.

Answers and Explanations

ANTIGENS

5.1. E. All are factors determining antigenicity. A hapten is a small foreign molecule too small to elicit the Ab response to itself unless bound to a larger carrier molecule, in which case, Ab response will be elicited to both carrier and hapten. Once Ab is formed in this manner, hapten can bind to Ab. Antigen, by definition, is capable of both eliciting and binding with the Ab. Also, by definition Ag and haptens would be foreign to the host.

5.2. C. Both Ag and hapten can bind with preformed Ab.

5.3. A. Ag can elicit Ab per se, but hapten cannot.

5.4. C. Both are foreign, by definition.

Antibodies—Immunoglobulins

Collectively, Antibodies

–are a heterogeneous group of proteins.
–consist of polypeptide chains linked by disulfide bonds.
–contain carbohydrate.
–have sedimentation coefficients ranging from 7S to 19S.
–are found predominately in the γ globulin fraction; some Abs are also in β and α fractions.

Reduction of Antibody Disulfide Bonds (Edelman)

–breaks S to S bonds between polypeptide chains.
–produces two identical heavy (H) chains and two identical light (L) polypeptide chains per Ab unit.
–causes cessation of Ab binding activity.
–show that H chains have MW $\approx$ 50K daltons.
–show that L chains have MW $\approx$ 25K daltons.

Papain Treatment of Immunoglobulin (Porter)

–produces two identical antigen-binding fragments (Fab) per Ig molecule.
–produces one crystallizable fragment (Fc) per Ig molecule.
–show that Fab fragments had univalent binding capability (i.e., would bind to Ag but could not precipitate Ag).

Pepsin Digestion of Immunoglobulin Molecule (Nisonoff)

–yield a large fragment that would precipitate Ag [(Fab')$_2$].
–show that (Fab')$_2$ fragments had bivalent binding capacity.

Antibody Molecules

–contain a minimum of two identical H and two identical L chains.
–have interchain disulfide bonds holding chains together, L to H, and H to H.
–have antigen-binding capacity defined by both H and L chains.

H Chains

–are polypeptide chains of 440 to 550 amino acid (a.a.) residues in length.
–have intrachain domains of approximately 110 a.a. residues formed by intrachain disulfide loops.
–have an amino terminal variable domain, followed by three to four constant domains.
–are structurally different for each of the defined classes of antibody (μ, γ, α, δ, and ϵ).

L Chains

–are polypeptide chains of approximately 220 amino acid residues in length.
–have intrachain domains of approximately 110 a.a. residues formed by intrachain disulfide loops.
–have an amino terminal variable domain and a carboxy-terminal constant domain.
–have two structurally distinct classes, kappa (κ) chains and lambda (λ) chains.

Variable Domains

–exist in both H and L chains.
–contain the hypervariable region (complementarity-determining region-CDR).
–are involved in antigen specificity of antibody (PARATOPE).

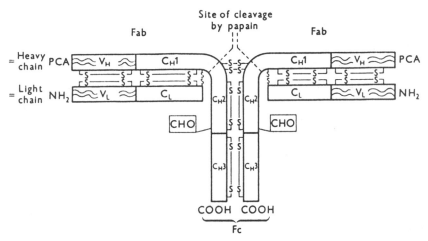

Figure 5.1 Diagram of the arrangement of peptide chains and disulphide linkages in the IgG_1 molecule. PCA = pyrrolidone carboxylic acid. NH_2 = amino terminal. COOH = carboxyterminal. CHO = carbohydrate. The site of cleavage by papain into Fab and Fc fragments is indicated.
V_H and V_L are the homologous variable regions of the heavy and light chains.
C_H1, C_H2, C_H3 = homology regions from constant region of heavy chain.
C_L = constant region from light chain (homologous to C_H1, C_H2 and C_H3). (After Edelman 1971.) (From Wilson GS, Miles A, eds. Topley and Wilson's Principles of Bacteriology, Virology and Immunity. 6th ed. Baltimore: Williams & Wilkins, 1975, vol 2, p 1377.)

Hypervariable Regions

–exist in both H and L variable domains.
–appear aproximately at amino acid positions 25 to 35, 50 to 58, and 95 to 108.
–define the antibody *paratope* to antigen *epitope*.

Classes (Isotypes) of Antibody Molecule

–defined by heavy chain characteristics:
γ, μ, α, ϵ, and δ
gamma, mu, alpha, epsilon, and delta.
–consist of two identical H chains and two identical L chains.
–are found in all species and all individuals.

Immunoglobulin G (IgG)

–has a molecular formula of $\gamma_2\kappa_2$ or $\gamma_2\lambda_2$.
–constitutes 73% of the Ig in serum on average.
–is referred to as 7S Ab.
–has a MW of 150,000 daltons.
–*crosses placenta* (except for IgG_4).
–*fixes complement* (not IgG_4).
–contains 3% carbohydrate.
–contains four subclasses (isotypes):
IgG_1-70% of IgG; IgG_2-19%; IgG_3-8% and IgG_4-3%, each of which has an antigenically and chemically distinct H chain defined mostly by amount and type of disulfide bridges.
–is the predominant Ab in the secondary or anamnestic response.

IgA

–molecular formula $\alpha_2\kappa_2$ or $\alpha_2\lambda_2$.

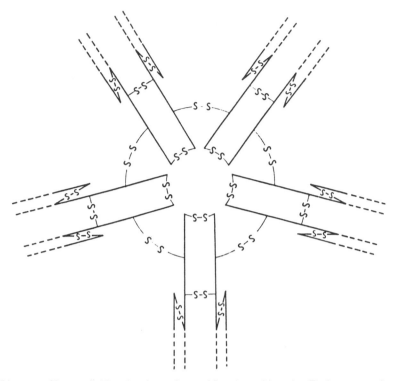

Figure 5.2. Diagram of human IgM molecule made up of five 4-peptide units. Each monomeric unit consists of a pair of μ heavy chains and a pair of light chains (λ or κ joined together by disulphide bonds. Each heavy chain has 4 interchain disulphide bonds; two between each pair of μ chains in the monomer, one to each light chain from a heavy chain, and one as a bridge between the monomers. The variable region (NH2-terminal) of each polypeptide chain is indicated by a dotted line, the constant region (COOH-terminal) by a continuous line. (From Wilson GS, Miles A, eds. Topley and Wilson's Principles of Bacteriology, Virology and Immunity. 6th ed. Baltimore: Williams & Wilkins, 1975, vol 2, p 1368.)

–constitutes 19% of total Ig on average.
–exists as monomer (7S) in serum, dimer (9S) in secretion.
–has a molecular weight of 160,000 daltons.
–contains 10% carbohydrate.
–is found in serum and in colostrum, respiratory and intestinal mucous membranes, saliva, and tears.
–consists of two subclasses, IgA₁ and IgA₂; the former is the predominant form in serum, and the latter is the predominant form in secretions. IgA₂ has no covalent bonds between L and α₂ chains.
–may have a joining (J) polypeptide chain holding two molecules together.
–may contain a secretory (transport) piece synthesized in the local epithelium, which may help retard autodigestion.

Immunoglobulin M (IgM)

–is a pentamer.
–has the molecular formula $(\mu_2\kappa_2)_5$ or $(\mu_2\lambda_2)_5$.
–is referred to as a macroglobulin.
–constitutes 7% of the total serum Ig.
–is referred to as the 19S antibody.

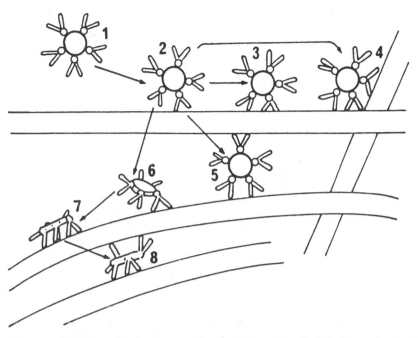

Figure 5.3. Diagram of the interaction between antiflagellar IgM and flagella. The free molecule (1) binds to a flagellum at a single point (2), then at more points (3). When another flagellum is within range, cross linking can take place (4, 5). The "staple" form is seen at (7). Its conversion from (2) is envisaged as occurring by means of an intermediate (6). A cross-linking "staple" form is shown at (8). (From a diagram kindly provided by Dr. A. Feinstein.) (From Wilson GS, Miles A, eds. Topley and Wilson's Principles of Bacteriology, Virology and Immunity. 6th ed. Baltimore: Williams & Wilkins, 1975, vol 2, p 1369.)

–has a molecular weight of 900,000 daltons.

–contains 15% carbohydrate.

–is the first immunoglobulin molecule to appear in ontogeny.

–is the first immunoglobulin to appear in response to Ag stimulation.

–is the first immunoglobulin to appear in phylogeny (lamprey eel has IgM).

–has a joining (J) polypeptide chain that holds the IgM pentamer together.

–has four constant domains on H chain: $C\mu 1$, $C\mu 2$, $C\mu 3$, and $C\mu 4$.

IgD

–has molecular formula $\delta_2\kappa_2$ or $\delta_2\lambda_2$.

–has a molecular weight of 150,000 daltons.

–constitutes 1% of serum immunoglobulin.

–has an unknown function in serum.

–serves as a receptor on the B-cell surface.

–contains 18% carbohydrate.

IgE

–has molecular formula $\epsilon_2\kappa_2$ or $\epsilon_2\lambda_2$.

–has molecular weight of 200,000 daltons.

–is referred to as *reaginic* antibody.

–is involved in allergic reactions.

–has four constant domains in the ϵ chain (three in Fc region, as in μ): $C\epsilon 1$, $C\epsilon 2$, $C\epsilon 3$, $C\epsilon 4$.

–has an Fc region that will bind to receptors on basophils and mast cells.
–contains 18% carbohydrate.

Allotypes of Immunoglobulins

–defined as small regular structural differences on molecules of Ig isotypes.
–may be determined by immunologic means.
–may be as simple as a single amino acid substitution on the H or L chains.
–are genetically defined and codominantly expressed.
–exist in IgG molecules as Gm_1-Gm_{20}.
–exist in IgA molecules as Am_1-Am_3.
–exist in κ chains as Km_1-Km_3.

Idiotype of Immunoglobulins

–is the antibody paratope to the antigenic epitope.
–is defined by the hypervariable regions (CDRs) of the variable domain of L and H chains.
–involves those determinants that define the binding capability of a given antibody.
–can be defined by immunologic means.
–number in the 10^6 range.

Idiotype-Anti-Idiotype Network

–is the concept that a given Ab idiotype can evoke the generation of an anti-idiotypic Ab.
–may control the generation and the level of the Ab response.

Review Test

ANTIBODIES

DIRECTIONS: For each of the questions or incomplete statements below, *one* or *more* of the answers of completions given is correct. Choose answer:

$$A = 1, 2, \text{ and } 3 \text{ correct}$$
$$B = 1 \text{ and } 3 \text{ correct}$$
$$C = 2 \text{ and } 4 \text{ correct}$$
$$D = 4 \text{ correct}$$
$$E = \text{all correct}$$

5.1. Antibody molecules:

1. Contain two identical H chains.
2. Contain carbohydrate.
3. Contain two identical L chains.
4. Are electrophoretically homogeneous.

DIRECTIONS: Each set of lettered headings below is followed by a list of numbered words or phrases. Choose answer.

Questions 5.2 through 5.7:
  A. IgA
  B. IgG
  C. IgM
  D. IgD
  E. IgE

5.2. The class of immunoglobulin important in protecting the mucosal surfaces of the respiratory, intestinal, and genitourinary tracts from pathogenic organisms.

5.3. Highest level in "normal" adult.
5.4. Highest level in "normal" 1-day-old child.
5.5. Implicated in atopic (i.e., allergic) response.
5.6. Initial Ig produced by B cells in response to Ag.
5.7. Longest half-life.

Answers and Explanations

ANTIBODIES

5.1. A. Ab molecules are composed, basically, of two identical H and two identical L chains held together by disulfide bonds. Each Ab molecule has a carbohydrate moiety associated with the H chain. Electrophoretically the Ab molecules are very heterogeneous, although the majority reside in the γ-globulin fraction.

5.2. A. IgA is the antibody associated with secretions and mucosal surfaces. Typically it would be an IgA dimer composed of IgA_2 subclass associated with a J chain and T piece.

5.3. B. IgG is the predominant Ig (approximately 72% of total Ig) found in the adult.

5.4. B. Although IgG is not made to any great extent in utero, IgG is readily passed transplacentally. The 1-day-old child would have adult levels of maternally derived IgG.

5.5. E. Allergic responses are the realm of IgE (reaginic) antibody.

5.6. C. The initial Ig produced in reaction to an Ag is IgM. IgM is also the first Ig to be synthesized in ontogeny.

5.7. B. IgG has the longest half life of the Igs (21-day average). The shortest half-life is IgE, approximately 48 hours.

Review Test

IMMUNOGLOBULIN CLASSES

DIRECTIONS: Each group of questions below consists of a series of lettered headings followed by a list of numbered words or phrases. Choose the *ONE* heading that is most closely related to the numbered words or phrases.

Questions 5.1 through 5.8

A. IgG
B. IgM
C. Both of the above
D. Neither of the above

5.1. Contain(s) λ chains.
5.2. Contain(s) α chains.

5.3. Encoded by ε exons.
5.4. Cross placental barriers.
5.5. J chain.
5.6. J gene.
5.7. Fc fragment.
5.8. S or T piece.

Answers and Explanations

IMMUNOGLOBULIN CLASSES

5.1. C. All of the immunoglobulin classes may contain either κ or λ Light (L) chains.

5.2. D. The heavy chain designates the isotype of Ig (α chain would thus be IgA, ε chain would be IgE, etc.).

5.3. D.

5.4. A. Of the Igs, only IgG crosses the placental barrier.

5.5. B. The J chain is a protein component holding α chains or μ chains together and thus is a component only of IgA and IgM.

5.6. C. The J gene exon encodes for a portion of the variable region of all L and H chains and could thus be said to be associated with all isotypes.

5.7. C. The Fc fragment is a product of papain cleavage of Ig—the other fragment being Fab. All of the Igs would be cleaved by papain and would render these two fragments.

5.8. D. The S (secretory) or T (transport) piece is associated only with IgA antibody found in secretions.

Immunoglobulin Genetics

Immunoglobulin Diversity

–is genetically determined.

–is independent of Ag availability.

–allows for the expression of upward of 10^6 different idiotypes.

–is generated by the selective rearrangement of immunoglobulin minigenes coding for L and H polypeptide chains.

Exons

–defined as minigenes or gene segments.

–encode for distinct parts of a polypeptide chain.

–L exon = codes for a leader segment.

  V exon = variable segment.

  D exon = diversity segment.

  J exon = joining segment.

  C exon = constant segment.

Introns

–intervening sequences between exons.

–spliced out via m-RNA transcription.

Formation of a Light Chain

–is genetically determined on human chromosome #2 (κ) or #22 (λ).

–involves the selection, at the germ line level, of genes for the leader sequence (L), variable (V) region, J (joining) segment, and constant (C) regions.

–is preceded by DNA rearrangement and deletion of the minigenes (exons) at the germ line level, and RNA splicing to remove intervening sequences (introns) in translation for the secreted light chain.

Formation of H Chain

–is genetically determined on human chromosome #14.

–involves the selection, at the germ line or embryonic DNA level, of 1 of 300 L and V genes, 1 of 4 J segments, and 1 of 12 D clusters.

–as in L chains, is preceded by DNA rearrangement and deletion of minigenes, and RNA splicing to remove introns before translation into the completed H chain.

Formation of Isotype

–is defined by the expression of one of the C exons encoding for the μ, δ, $\gamma3$, $\gamma1$, $\alpha1$, $\gamma2$, $\gamma4$, ϵ, or $\alpha2$, in that order, on chromosome #14.

–follows the determination of idiotype, i.e., assembly of the VDJ exons.

Immunoglobulin Assembly

–is initiated by the formation of a functional L chain from either chromosome #2 (κ) or chromosome #22 (λ).

–is followed by the assembly of a specific H chain isotype.

–involves the joining of H chains and the joining of L to H chains by disulfide bonding and the secretion of the completed molecule.

Antibody Diversity

–is generated by the random assembly of light chain V and J genes.

–is generated by the random assembly of heavy chain V, D, and J genes.

–is generated by the random assembly of H and L chains.
–is due to errors in recombination of the V, D, and J genes.
–is due to somatic mutations.

Review Test
IMMUNOGLOBULIN GENETICS

DIRECTIONS: Each group of questions below consists of a series of lettered headings followed by a list of numbered words or phrases. Choose the *ONE* heading that is most closely related to the numbered words or phrases.

Questions 5.1 through 5.5
  A. Variable domain of H chain
  B. Constant domain of H chain
  C. Both of the above
  D. Neither of the above
5.1. J gene.

5.2. D gene.
5.3. L gene.
5.4. Mu (μ) gene exons.
5.5. Fc fragment.

Answers and Explanations

IMMUNOGLOBULIN GENETICS

5.1. A. The variable region of the H chain is encoded by the rearranged V, D, and J exons located on chromosome # 14.

5.2. A. See #1.

5.3. D. The L gene exon encodes for the leader sequence, which is cleaved before the secretion of the completed Ig molecule.

5.4. B. The μ exon encodes the constant domains of IgM.

5.5. B. The Fc fragment is a product of papain cleavage of Ig and consists of the C terminal constant domains of the H chain.

Antigen-Antibody Reactions

Forces Holding Ag-Ab Together

–are identical to any protein-protein interaction.
–are *NOT* covalent.
–include the following:

(a) Ionic - attraction of oppositely charged groups
   (e.g., NH_3+ . . . . . -OOC)
(b) Hydrogen bonding - sharing of hydrogen by hydrophilic groups.
(c) Hydrophobic interaction - exclusion of water by valine, leucine, phenylalanine.
(d) Van der Waals forces - weak magnetic field.

Affinity

–is the tendency to form a stable complex.
–applies to a specific Ab against a specified EPITOPE (i.e., a single Ab-Ag reaction).
–is defined by the formula:

$$\text{Affinity} = \frac{k}{k'} = \frac{[SL]}{[S]\,[L]}$$

where

$$
\begin{aligned}
k &= \text{association constant} \\
k' &= \text{disassociation constant} \\
S &= \text{binding site of Ab (PARATOPE)} \\
L &= \text{ligand (EPITOPE)} \\
[\,] &= \text{concentration}
\end{aligned}
$$

which is derived from the standard protein (enzyme) formula:

$$S + L = SL$$

Avidity

–is the sum of the affinities.
–refers to the total Ab response to the whole of the epitopes associated with a given antigen.

Quantitative Precipitin Curve

–is generated by a series of reactions involving (usually) a constant Ab amount titrated against increasing amounts of Ag.
–for example:

–nine test tubes are prepared, each with 0.1 ml *antiserum* to ovalbumin (i.e., anti-OA Ab).
–each tube receives increasing amounts of Ag (tube #1 = 0 μg OA, #2 = 10 μg OA, #3 = 20 μg OA, ... , tube #9 = 80 μg OA).
–on incubation one should be able to visualize a *zone of Ab excess (tubes 1 to 3)*, a *zone of Equivalence* (tubes 4 to 6), and a *zone of Ag excess* (tubes 7 to 9).

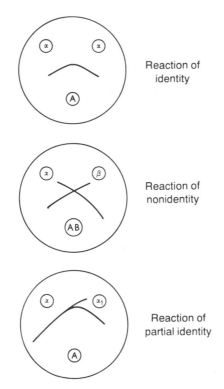

Reaction of identity

Reaction of nonidentity

Reaction of partial identity

Figure 5.4. Double diffusion in two dimensions (Ouchterlony procedure). α and β are two unrelated antigens. A and B are antisera prepared against α and β. α_1 is an antigen that cross-reacts with α. (From Wilson GS, Miles A, eds. Topley and Wilson's Principles of Bacteriology, Virology and Immunity. 6th ed. Baltimore: Williams & Wilkins, 1975, vol 1, p 299.)

–the amount of precipitation of Ag to Ab will be greatest at the *zone of Equivalence.*

Lattice Theory

–states that precipitate will form in a lattice arrangement under optimum relationships of Ab to Ag.
–Ab or Ag excess will diminish a lattice network and decrease amount of precipitate.

Ring Test

–a precipitation reaction that takes place at the interface between two solutions, one containing Ag, one containing Ab.

Oudin: Single Diffusion

–a precipitation technique, usually accomplished in a *test tube,* where Ab (or Ag) in a gel is allowed to react with soluble Ag (or Ab) diffusing through it from a liquid interface.

Ouchterlony: Double Diffusion

–a precipitation technique in agar, usually accomplished in a *petri dish,* where Ag and Ab are allowed to diffuse against each other and permit the formation of a precipitin line between the sample wells.
–if two or more sample (Ag) wells are used to a single antiserum (Ab) well one can distinguish:

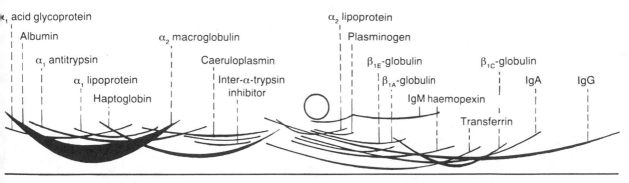

Figure 5.5. Electrophoretic pattern obtained with human serum. (From Wilson GS, Miles A, eds. Topley and Wilson's Principles of Bacteriology, Virology and Immunity. 6th ed. Baltimore: Williams & Wilkins, 1975, vol 1, p 303.)

–if two or more sample (Ag) wells are used to a single antiserum (Ab) well one can distinguish:

Lines of identity -a single precipitin line indicating uniformity and identity of the Ags in the sample well.

Lines of nonidentity - two distinct and crossing precipitin lines indicating no cross-reactivity or identity between two Ag sample wells.

Lines of partial identity - a *spur* formation, indicating a sharing of identity between two Ag wells. This would indicate that some, but not all of the antigenic determinants (epitopes) are shared between two Ags.

Immunoelectrophoresis (IEP)

–an Ag-Ab reaction technique in agar that combines *electrophoresis* [i.e., separation of serum proteins at a given pH (usually pH = 8.4) with an electric current to their constituent parts: albumin, α_1, α_2, β_1, β_2, and γ-globulin fractions] and then a precipitin reaction of these separated proteins to specific Ab placed in the antiserum trough.

Rocket Immunoelectrophoresis

–a rapid method for estimating Ag concentration.
–sample wells are punched at one end of a gel plate in which Ab to specific Ag has been dissolved.
–samples are applied and electrophoresed.
–rocket-like precipitate lines will form; the length will depend on initial Ag sample concentration, and a reference line is determined.
–unknown concentration is determined by length of arc relative to standard reference.

Radial Immunodiffusion

–method for estimating Ag concentration.
–Ab to specific Ag is diffused in agar gel or slab.
–Ag wells are cut and samples of various known concentrations applied.
–gel is incubated and precipitin rings developed.
–diameter of precipitin ring is proportional to initial concentration; reference line is determined.
–unknown sample is determined by comparison to reference line.

Radioimmunoassay (RIA)

–a very sensitive assay for Ag that uses (usually) a known amount of labelled Ag (Ag*) and a known amount of specific Ab for that Ag.

–A standard inhibition ("quench") curve is then generated by reacting increasing known amounts of unlabeled Ag with the constant Ab amount and determining the Bound Ag*-Ab to Free Ag* ratio. (KEY: The more unlabeled Ag present the less Ag* will be bound to Ab.)

–unknown sample is quantified by interpolation of the standard inhibition curve.

Radioimmunosorbent Test (RIST)

–RIA assay for total IgE.

–*Procedure:* solid phase surface (dextran beads, e.g.) coupled with anti-IgE Ab.

–a known concentration of labeled IgE* is reacted with this bound Ab with increasing known concentrations of unlabeled IgE to produce a *standard inhibition curve.*

–unknown serum sample is then tested.

–concentration of unknown determined by extrapolation of standard inhibition curve.

Radioallergosorbent Test (RAST)

–a RIA test to determine *specific* IgE concentration.

Procedure

–solid phase beads are coupled with specific Ag.

–known amounts of specific IgE are reacted with bound Ag.

–this in turn, reacted with labeled anti-IgE*.

–a standard curve is generated.

–unknown serum sample (containing putative IgE to specific Ag) is reacted with beads.

–labeled anti-IgE (anti-IgE*) is reacted with the Ag-IgE on bead.

–the concentration of specific IgE determined from the standard curve.

ELISA (Enzyme-Linked Immunosorbent Assay)

–same test as RIA (RIST and RAST) except *enzyme* is attached instead of a radioactive label and different Ig are detected (not just IgE).

–enzyme will cause a color change on the addition of a specific substrate.

–intensity of color change will be proportional to amount of reactivity.

Immunocompetent Cells in the Immune Response

Cells Involved in the Elicitation of the Immune Response

–Macrophages, monocytes, or antigen presenting cells (APC).
–T cells.
–B cells.

Antigen Processing Cells

–include macrophages, monocytes, and their derivatives, including microglial cells, Kupffer's cells and skin Langerhans' cells.
–are characterized by dendridic extensions and ability to phagocytize, internalize, and "process" antigen.
–possess Ia antigen, Fc receptors, and C3b receptors.
–produce Interleukin I.

T Cells

–are thymus-dependent lymphocytes.
–mature and develop in the thymus.
–have a unique antigen receptor of a specific idiotype (Ti).
–develop a series of thymus-induced differentiation markers labeled as Clusters of Differentiation (CD).
–have Fc receptors on some subsets and C3b receptors.

Ti

–is the antigen-specific (idiotype) receptor on T cells.
–consists of a heterodimer.
–α chain is 43K, β chain is 49K; each has two external domains, a transmembrane segment and a cytoplasmic extension.
–α chain encoded on chromosome #7; β chain on chromosome #14.
–encoded by a DNA rearrangement of V, D, and J exons for the V region and a C gene for the C region.
–associated with CD3.

CD Markers on T Cells

–arise on T cells during maturation sequence in thymus.
–appear on T cells in the following sequence:

CD2 (T11)
CD3 (T3)
CD4 and CD8 (T4 and T8).

CD2

–earliest T-cell marker.
–is the SRBC receptor.
–probably a primative or generalized T-cell receptor.
–present on virtually every peripheral T cell.
–50K polypeptide.

CD3

–intimately associated with Ti.
–composed of three molecules (19K, 22K, 25K), essentially transmembrane and cytoplasmic, that transduce signals from Ti.

CD4

–present mainly on *T Helper Cells.*
–involved in interaction with *class II* HLA.
–molecular weight 62K.

CD8

–present mainly on *T cytotoxic and T suppressor cells.*
–recognize *class I HLA.*
–molecular weight 76K.

Ontogeny of T Cells

–occurs as stem cells flow through thymic cortex, into medulla and then out into general circulation.
–begins in cortex with appearance of CD2, followed by the appearance of CD3 (with Ti), then with concomitant expression of CD4 & CD8.
–in medulla, consists of a loss of marker to produce two populations of cells, one CD2+, CD3+, Ti+, CD4+ (65%) and the other CD2+, CD3+, Ti+, CD8+ (35%), which are then released into the peripheral circulation.
–is also time when "self-recognition" via HLA markers occurs.

B Cells

–are thymus-independent lymphocytes.
–arise from stem cells.
–develop independent of antigen.
–mature in bone marrow (bursa of Fabricius in birds) and/or bursal equivalent.
–have a unique surface immunoglobulin (S-Ig) receptor for antigen.
–develop a series of markers during the differentiation process.

S-Ig

–is the antigen-specific idiotype receptor on B cells.
–is equivalent to an antibody molecule with a transmembrane projection.
–contains two H chains and two L chains.
–H chain encoded on chromosome #14 and thus may be μ, δ, $\gamma3$, $\gamma1$, α, $\gamma2$, $\gamma4$, ϵ, or $\alpha2$.
–L chain encoded on either chromosome #2 (κ chain) or on #22 (λ chain).
–V region encoded by DNA rearrangement of V and J genes for L chain, and V, D, and J genes for H chain.
–exists in 100,000 copies/mature B cell.
–will undergo "capping" and endocytosis following activation by Ag.

Ontogeny of B Cells

–the process by which a stem cell undergoes differentiation from a pre-B cell, to an immature B cell, to a mature B cell is driven by Ag to an activated B cell, and finally to a plasma cell capable of producing immunoglobulin.
–is initiated in the fetal liver and adult bone marrow and would usually terminate in the germinal centers of lymph nodes and spleen.

Sequential Appearance of B-Cell Markers

–stem cell—HLA class I and class II (HLA-DR) antigen.
–pre-B cell—cytoplasmic μ chains.

–immature B cell—membrane-bound IgM subunits, C3b, Fc, and Epstein virus receptors.

–mature B cell—membrane-bound IgM, IgD, IgG, IgA, IgE.

–activated B cell—"capping" of S-Ig.

–plasma cell—few S-Ig and DR molecules; NO Fc, C3b, and E-B virus receptors.

Cell Interactions in the Immune Response

Macrophages/Antigen-Presenting Cells

–phagocytize the antigen.
–process and degrade the antigen.
–express antigenic determinants (epitopes) of the antigen on its surface in the context of HLA class II (Ia, DR) molecules.
–produce Interleukin I and numerous other monokines.

Interleukin I (IL-1)

–produced by many different types of cells; relatively high concentrations are produced by macrophages, monocytes, skin Langerhans' cells and other dendritic cells.
–formerly known as Lymphocyte Activating Factor (LAF).
–augments activity of many cell types, especially T cells.
–is a peptide of 17,500 daltons.
–is heat and pH stable.
–is endogenous pyrogen (EP).
–induces increase in acute phase reactants.
–occurs in two forms, IL-1 α and β.

T Helper Cells

–recognize antigen epitopes in the context of the class II molecules by Ti and CD4 respectively.
–the antigenic signal then is transduced internally via the CD3 molecules.
–react with IL-1 from APC.
–produce IL-2.
–express IL-2 receptors.
–become activated when IL-2 receptors are filled with IL-2.
–elaborate B-cell growth factor.
–elaborate B-cell differentiation factor.
–elaborate other lymphokines.
–elaborate interferon γ.

Interleukin II (IL-2)

–produced by T lymphocytes and *large granular lymphocytes* (LGL).
–formerly known as *T-cell growth factor* (TCGF).
–augments proliferation of T (and B?) cells.
–enhances activity of T cells and NK (natural killer) cells.
–is a glycoprotein of 15,500 daltons.
–is heat labile.

Interferon Gamma (γINF)

–produced by activated T cells and LGL.
–increases expression of Ia on macrophages, B cells, and other APC.
–molecular weight of 17,000 daltons.
–heat labile.
–has antiviral properties.
–has regulatory control in the immune response.

B-Cell Growth Factor (BCGF) (IL-4)

–produced by activated T cells.

–induces antigen-stimulated B cells to proliferate (blast transformation).

B-Cell Differentiation Factor (BCDF) (IL-5)

–produced by activated T cell.
–stimulates blasting B cell to differentiate into antibody-producing plasma cells.

B Cells

–recognize epitopes via surface immunoglobulin receptors.
–cross-link such epitopes on the B-cell surface.
–surface receptors undergo capping and internalization by pinocytosis.
–are stimulated to blastogenesis by BCGF.
–are induced to differentiate into plasma cells by BCDF.

T-Independent Antigens

–do not need T-helper activity, T-cell activation, or the production of BCGF and BCDF.
–are polymeric in nature.
–examples are endotoxin, lipopolysaccharide, polyvinylpyrrolidone, polymerized flagellin, E-B virus, dextran.
–produce a primary (IgM) response.
–do not produce memory cells or an anamnestic, secondary (IgG) response.

T-Cell Mitogens

–polyclonal activators of T cells.
–examples are concanavalin A (Con A) and phytohemagglutinin (PHA).
–stimulate general T-cell activation by cross-linking specific sugars on the cell surface.

B-Cell Mitogens

–polyclonal activators of B cells.
–examples are lipopolysaccharide, endotoxin, polyvinylpyrrolidone, polymerized flagellin, E-B virus, dextran.

Review Test
CELL-TO-CELL INTERACTION

DIRECTIONS: Each group of questions below consists of a series of lettered headings followed by a list of numbered words or phrases. Choose the ONE heading that is most closely related to the numbered words or phrase.

Questions 5.1 through 5.5
A. $CD4^+$ cell
B. $S\text{-}Ig^+$ cell
C. Both of the above
D. Neither of the above

5.1. Produce(s) BCGF.
5.2. Produce(s) BCDF.
5.3. Produce(s) IL-1.
5.4. Produce(s) IL-2.
5.5. Produces IFN γ.

Questions 5.6 through 5.9
A. T cells
B. B cells
C. Both of the above
D. Neither of the above

5.6. Possess specific antigen receptor.
5.7. Have Class I histocompatibility Ags on surface.
5.8. C3b receptor.
5.9. Have Class II histocompatibility Ags on surface.

Questions 5.10 through 5.12
A. CD2 antigen
B. CD8 antigen
C. Both of the above
D. Neither of the above

5.10. T_H (T helper) cell.
5.11. T_S (T suppressor) cell.

5.12. T_C (T cytotoxic) cell.

Questions 5.13 through 5.15
A. T-cell mitogen
B. B-cell mitogen
C. Both of the above
D. Neither of the above

5.13. Phytohemagglutinin.
5.14. Pokeweed mitogen.
5.15. Bacterial lipopolysaccharide.

Questions 5.16 through 5.18
A. Interleukin 2 (IL-2)
B. γ Interferon (IFNγ)
C. Both of the above
D. Neither of the above

5.16. Produced by CD2 cells.
5.17. Specific receptor present on NK and K cells.
5.18. Has effect on macrophage.

Questions 5.19 through 5.20
A. Macrophage Migration Inhibition Factor (MIF)
B. Interleukin I (IL-I)
C. Both of the above
D. Neither of the above

5.19. Made by macrophages/monocytes.
5.20. Endogenous pyrogen.

Answers and Explanations

CELL-TO-CELL INTERACTION

5.1. A. B-cell growth factor (BCGF = IL-4) is a product of the activated T helper (CD4$^+$) cell.

5.2. A. B-cell differentiation factor (BCDF) i.e., also a product of the activated T helper (CD4$^+$) cell.

5.3. D. Interleukin-1 (IL-1) is produced by macrophages-monocytes and activates lymphocytes, especially CD4$^+$ cells, to produce IL-2.

5.4. A. IL-2 is a product of activated T helper cells (CD4$^+$ cells). It causes T cells to undergo differentiation and full activation. IL-2 also causes an increase in IL-2 receptor formation on the surface of CD4 cells.

5.5. A. Gamma interferon is produced by antigen-stimulated CD4 cells. γIFN is a powerful immunomodulator.

5.6. C. Both T and B cells possess a specific antigen receptor. The T-cell receptor is a dimer; the B-cell receptor is a membrane-bound Ig molecule. Both have variable and constant regions and similar genetic development.

5.7. C. The Class I histocompatibility Ags are present on virtually every nucleated cell in the body. Both T and B cells would have these Ags.

5.8. C. The C3 receptor appears to be a ubiquitous structure on almost all immunocompetent cells: monocyte/macrophages, neutrophils, lymphocytes, etc.

5.9. B. The Class II Ags appear mostly on B cells and on macrophages/monocytes. These Ags are "recognized" by the CD4 molecule present on T helper cells.

5.10. A. The T_H cell would possess CD2, CD3, and CD4 Ags.

5.11. C. The T_S cells would possess CD2, CD3, CD8 Ags.

5.12. C. The T_C cell is also CD2$^+$, CD3$^+$, and CD8$^+$.

5.13. A. PHA is a T-cell mitogen.

5.14. C. PWM is a mitogen for B *and* T cells.

5.15. B. LPS is a B-cell mitogen.

5.16. C. The CD2 marker is the sheep RBC receptor ubiquitous on T cells. IL-2 and γIFN are produced by the fully activated CD4$^+$ cell, which could also have the CD2 marker.

5.17. C. Both K and NK cells are stimulated to activation both by IL-2 and γIFN and have specific receptors for these compounds.

5.18. B. γIFN has a specific effect on macrophages, causing an increase in the expression of Ia (Class II histocompatibility) antigens on the surface.

5.19. B. The macrophage manufactures IL-1, which causes T helper cells to be activated. Macrophage MIF is a product of activated T helper cells.

5.20. B. Endogenous pyrogen and IL-1 are identical molecules.

Immunology of AIDS

AIDS

–is *a*cquired *i*mmuno*d*eficiency *s*yndrome.

–is caused by human immunodeficiency virus I (HIV-1).

–is characterized by the loss of T helper (CD4$^+$) cell population.

–allows for an increase in opportunistic infections (e.g., *Pneumocystis carinii*), neo-plasms (e.g., Kaposi's sarcoma, a tumor of blood vessel tissue of skin and internal organs), skin disorders, and neurologic disease.

Brief History of AIDS

–current epidemic known since 1981, probably starting in central Africa and going to U.S. and Europe via Caribbean Islands.

–causative agent identified by Luc Montagnier of the Pasteur Institute as *l*ymphade-nopathy-*a*ssociated *v*irus (LAV) in 1983 and by Robert Gallo on the NIH as human T lymphotropic virus-III (HTLV-III) in 1984. Virus is now referred to as HIV-I by international court settlement.

–test for anti-HIV-1 antibody available as of 1985.

AIDS Testing

–primary test is for presence of anti-HIV-I antibody (ELISA test).

–confirming test is by *Western Blot* techniques: HIV proteins are electrophoresed, putative serum antibody of patient reacted and read by antihuman antibody conjugated with enzyme or radioactive label.

Prevalence of AIDS by Group

82% Homosexual/bisexual males
4% IV drug users
6% Homosexual/bisexual and IV drugs
3% Hemophilia
4% Heterosexual transmission
1% Transfusion
1% Undetermined

Projection for 1991 is 270,000 AIDS patients in U.S.

HIV Infection Methodology

–sexual contact.

–blood and blood products.

–perinatally, mother to infant.

–most effective via whole infected cells.

–drug abusers via shared needles.

HIV Virus

–is an RNA retrovirus.

–member of the *Lentivirus* family of retroviruses.

–has an external membrane derived from host cell.

–has glycoproteins on and in the membrane (gp120 and gp41) encoded by *env* gene.

–has protein core (p18 and p24) encoded by *gag* genes.

–has two copies of the viral RNA genome and the reverse transcriptase encoded by the *pol* gene within the core.

–has a proviral genome of 10kb with *long terminal redundancies* (LTR) at each end; the *gag, pol,* and *env* genes; and at least five regulatory genes.

Effect of HIV Infection on the Immune System

–HIV-1 destroys $CD4^+$ (T helper) cells, which are pivotal in the immune response.

–receptor for HIV is the CD4 molecule probably via gp120 and gp41.

–after binding, the virus is pinocytozed and uncoated.

–RNA is transcribed to DNA by reverse transcriptase (*pol* gene product) and is inserted into human chromosomal DNA by endonuclease (*pol* product) as the *PROVIRION*.

–provirus is latent stage.

–activation of cell causes DNA → RNA → m-RNA transcription → protein synthesis → virus → virus budding with destruction of $CD4^+$ cell.

–death of cell may be due to formation of *syncytia* by infected cell gp120 interaction with CD4 of noninfected cells.

–HIV envelope proteins may mimic parts of class II HLA antigens, leading to binding HIV to CD4 and loss of immune response.

–macrophage and some brain cells may have low levels of CD4 on surface, allowing for tropism by HIV but not destruction of carrier cells.

Treatment for Virus Infection

–difficult because of nature of virus.

–vaccine ineffective to date due to extensive mutability of virus envelope proteins gp120 and gp41.

–chemical treatment consists chiefly (as of 1988) of azidothymidine (AZT), which blocks viral RNA → DNA transcription.

Histocompatibility Antigens

The Major Histocompatibility Complex (MHC)

–is located on the short (p) arm of human chromosome #6.

–encodes for the major human histocompatibility antigens, known collectively as HLA (human leukocyte antigens).

–consists of genes encoding for HLA-A, HLA-C, HLA-B, HLA-DR, HLA-DQ, and HLA-DP antigens as well as some complement components.

–encodes for class I HLAs by the genes HLA-A, HLA-B, and HLA-C.

–encodes for class II HLAs—defined by the genes within the D (immune response) region: HLA-DR, HLA-DP, HLA-DQ.

–encodes for class III HLA antigens—defined by the genes encoding for various complement components, including C2, C4, factor B, C3b receptor, etc.

The Class I HLA Antigens

–are encoded by HLA-A, HLA-B, and HLA-C genes.

–are expressed on virtually every nucleated cell in the body.

–are associated noncovalently with a 12K polypeptide (B2 microglobulin), which is encoded on chromosome #2.

–consist of a 44K dalton polypeptide chain containing transmembrane and cytoplasmic components.

–have three external domains analogous to Ig domains (B2 microglobulin defines the fourth domain).

–are involved in recognition (restriction) by CYTOTOXIC T cells, bearing appropriate receptors for the HLA antigens (CD8).

–are associated with the expression of antigen on virus-infected or transformed cells.

The Class II Antigens

–are encoded by genes in the HLA-D region: HLA-DR, HLA-DQ, and HLA-DP.

–are expressed mainly on B cells, macrophages, and other immunocompetent cells.

–consist of an α chain (29K) and β chain (34K) heterodimer, with cytoplasmic and transmembrane components.

–possess two external domains on each of the polypeptide chains.

–allow interaction (restriction) between antigen-presenting cells (APC) and T cells via CD4.

–are associated with the presentation of Ag on antigen-processing cells.

Review Test

HISTOCOMPATIBILITY ANTIGENS

DIRECTIONS: Each group of questions below consists of a series of lettered headings followed by a list of numbered words or phrases. Choose the ONE heading that is most closely related to the numbered words or phrase.

Questions 5.1 through 5.4
A. Class I histocompatibility Ag (HLA-A)
B. Class II histocompatibility Ag (HLA-DR)
C. Both of the above
D. Neither of the above

5.1. CD4 cell is "restricted" to seeing Ag in this context.

5.2. CD8 cell is "restricted" to seeing Ag in this context.

5.3. Present on B cells.

5.4. Involved in ADCC (antibody-dependent cell cytotoxicity).

Answers and Explanations

HISTOCOMPATIBILITY ANTIGENS

5.1. B. The CD4 molecule is the receptor for the Class II (Ia) antigens. The CD4 cell (helper T cell) is restricted in seeing Ag in the context of the Ia antigens.

5.2. A. The CD8 molecule is the "receptor" for the Class I Ag. Cytotoxic T cells (CD8$^+$) see viral Ag in the context of the Class I Ag.

5.3. C. B cells have on their surface both the Class I Ags (which are present on the surface of *all* nucleated cells of the body) and Class II Ags (present on B cells and macrophages).

5.4. D. Antibody-dependent cell cytotoxity is a function of K cells and is dependent on antibody coating of a target cell. The histocompatibility Ags have no known function in ADCC.

Tumor Immunology

Tumor Cells

–are cells transformed by virus or by chemical or physical means.
–are usually not well differentiated.
–lack contact inhibition and usually proliferate uncontrollably.
–have *tumor associated antigen* (TAA) expression on surface.

Viral Tumors

–caused by RNA and/or DNA oncogenic viruses.
–share TAA epitopes on cells transformed by same or similar viruses and may have cross-reactivity.

Chemically Induced Tumors

–caused by a variety of chemical carcinogens (e.g., coal tar, methylcholanthrene).
–share virtually no TAA epitopes and have no cross-reactivity.

Physically Induced Tumors

–caused by physical "insult" such as UV light, x-ray or γ-radiation.
–behave akin to chemically induced tumors, i.e., no cross-reactivity.

Oncofetal Antigens

–antigenic epitopes found in normal fetal development and on transformed (tumor) cells.
–are not usually present on normal adult cells.

Examples of Oncofetal Ags

Carcinoembryonic Ag (CEA)

–normally found in gut, liver, and pancreas in months 2 to 6 of gestation.
–appears in 90% of pancreatic Ca's, 70% of colon Ca, 35% of breast Ca, and 5% of normals.
–increases during pregnancy and in recurrence of tumor.

α-Fetoprotein (AFP)

–normal α-globulin of embryonic/fetal serum.
–appears in hepatomas, in cirrhosis, and in hepatitis.

Immune Mechanisms in Tumor Cell Destruction

–TAA will elicit both the *humoral* and the *cell-mediated* mechanisms.
–*antibody binding* to tumor cells will kill tumors by:

1. attachment of Ab to Fc receptors on macrophages, PMNs and subsequent *phagocytosis*.
2. attachment of Ab to Fc receptors on null (killer) cells and subsequent lysis by antibody dependent cell cytotoxicity (ADCC).
3. activation of the complete complement sequence and causing tumor *lysis*.
4. activation of the complement sequence to produce C3b on tumor cell surface that will react with C3b receptors on macrophages and PMNs to enhance *phagocytosis*.

–*Cell-mediated immunity* to tumors includes:

1. Production of lymphokines by T helper cells to mobilize and activate *macrophages* against tumor cells. These would include *m*acrophage *a*ctivating *f*actor, macrophage migration inhibition factor, macrophage chemotactic factor.
2. Production of monokines by *activated macrophages*—i.e., *t*umor *n*ecrosis *f*actor (TNF).
3. *Activation* of $CD8^+$ *cytotoxic T cells* that recognize TAA on virally transformed cell Ags in association with Class I histocompatibility Ags.
4. Killing of tumor cells spontaneously and without previous sensitization by natural-killer (NK) cells.

Blood Group Immunology

the ABOH System
–defined by the presence of alloantigens A and B on RBC surface.
–determined by isohemagglutinins (predominately IgM) in circulation.
–define A, B, O, and AB individuals.
–determined genetically by codominant A, B, and O genes.

ABOH Blood Types
–Type (Phenotype) *A* has A alloantigen on RBC and anti-B isohemagglutinin in serum. Genotype could be AA (homozygous) or AO (heterozygous).
–Type *B* has B alloantigen on RBC and anti-A isohemagglutinin in serum. Genotype would be BB or BO.
–Type *AB* has both A and B alloantigens on RBC and no isohemagglutinins. Genotype is AB.
–Type *O* has neither A nor B alloantigens but has both anti-A and anti-B isohemagglutinins. Genotype is OO. O gene is a silent (amorphic) gene.

Rh Antigens
–determined by complex genes DCE/dce of which D antigen is most important medically. DCE/dce has codominant expression.
–are Rh positive if D gene expressed, Rh negative if D gene not expressed.

Rh Incompatibility
–is important in pregnancy.
–can lead to hemolytic disease of newborns.
–can occur only in case where mother is Rh negative and fetus is Rh positive.
–is caused by "spill" of large numbers of Rh positive RBC into maternal circulation during birth of first Rh positive infant. Mother will be sensitized and produce high levels of anti-Rh IgG.
–subsequent Rh positive fetuses can be "attacked" by maternal anti-Rh IgG and "erythroblastosis fetalis" can occur.
–can be prevented by treatment of mother with anti-Rh antibody within 72 hours of each Rh positive delivery. This treatment will destroy the fetal RBC, prevent maternal sensitization, and eliminate the production of anti-Rh antibody by mother's immune system.

Review Test

BLOOD GROUPS

DIRECTIONS: For each of the questions or incomplete statements below, *one* or *more* of the answers or completions given is correct. Choose answer:

$$A = 1, 2, \text{ and } 3$$
$$B = 1 \text{ and } 3$$
$$C = 2 \text{ and } 4$$
$$D = 4$$
$$E = \text{all}$$

5.1. Anti-A isohemagglutinins are present in persons with:

1. Type A blood.
2. Type B blood.
3. Type AB blood.
4. Type O blood.

5.2. Substances that may be passively transferred from the mother to the fetus during the third trimester of pregnancy include

1. IgG.
2. IgM.
3. anti-Rh antibody.
4. natural isoagglutinins.

DIRECTIONS: Pick the *one* best answer.

5.3. A male heterozygous for Rh factor mates with an Rh negative female. On the basis of genetic theory it could be predicted that:

A. no offspring would be Rh positive.
B. 25% of offspring would be Rh positive.
C. 50% of offspring would be Rh positive.
D. 100% of offspring would be Rh positive.
E. no statement possible.

Answers and Explanations

BLOOD GROUPS

5.1. C. Type B individuals have anti-A antibody. Type O individuals have both anti-A and anti-B isohemagglutinins.

5.2. B. IgG is the only immunoglobulin that passes the placental barrier. Anti-Rh antibody of the IgG type will readily cross the placenta and cause erythroblastosis fetalis. IgM will not pass the placenta nor will the majority of natural isohemagglutinins (IgM molecules predominately).

5.3. C. A heterozygous individual, by definition, would be Dd relative to Rh. An Rh negative female would be described as dd. A Dd x dd cross would allow for 50% of the offspring to be Rh positive.

Complement

Complement-Mediated Cell Cytotoxicity

–causes lysis of a target cell.
–may be initiated by Ab fixation to a cell surface antigen.
–may be due to Ag-Ab complex formation.

Complement Components

–a collective term for a group of heterogeneous proteins involved in a sequential activation culminating in target cell lysis.
–are not immunoglobulins.
–are present in "normal" serum.
–do not increase as a result of Ag stimulation.
–are manufactured very early in ontogeny (first trimester).
–are made in macrophages and liver (except C1, which is made and assembled in GI epithelium).
–are heat labile.
–are defined by number (C1, C2, C3) and letter (Factor B, Factor D) designations.

Complement Activation Can Occur

–by the *classical* complement cascade—i.e., mediated by IgG, IgG2, IgG3, or IgM antibody.
–by the *alternate* pathway—i.e., initiated by certain antigens (i.e., lipopolysaccharide, endotoxin, zymosan) and Ag-Ab complexes.

Classical Activation Pathway Sequence

–a singlet of IgM or doublet of IgG1, IgG2, IgG3, very closely spaced on the cell surface, will bind C1 by the C1q region via a receptor in the Fc region ($C\mu 4$ or $C\gamma 3$).
–C1 consists of subunits C1q, C1r, and C1s bound by Ca^{++}.
–C1 will be activated from a proesterase to an esterase, cleaving both C4 and C2 sequentially.
–C4 will be cleaved into C4b and C4a, C4b attaching at site of Ab fixation, C4a going off in fluid phase and may act as an anaphylatoxin.
–C2 will be cleaved into C2a and C2b, C2a attaching to C4b via Mg^{++}. C2b is released in fluid phase, and acts as a kinin.
–Ab-C1-C4bC2a complex is C3 convertase.
–C3 is cleaved into C3a and C3b; C3b has receptors on many cell surfaces, including PMNs and Mϕ; C3a is Anaphylatoxin I and will attach preferentially to C3a receptors on mast cell and cause histamine release.
–C5 is analogous to C3 and is cleaved into C5b and C5a. *C5b* will cause activation of C6 and C7, is chemotactic for PMNs and Mϕ. *C5a* is Anaphylatoxin II and has receptors on mast cells for histamine release.
–C6 and C7 are activated by C5b and exist as an activated entity *C5b67,* which is chemotactic for PMNs and Mϕ.
–C8 and C9 will cause permeability changes and allow water influx: cell swells and lysis occurs.

Control Mechanisms of Complement Activation

Instability of Components

–C2a and C5b have a very short half-life.

Inhibitors

–C1 Inhibitor (C1-INH) stops continued activation of C1. Congenital lack of C1-INH is called *hereditary angioedema*.

Inactivators

–*C3b inactivator* is enzyme that destroys C3b activity. β*1H* enhances action of C3b-inactivator.

–*C6 inactivator*—destroys C6 activity.

–*anaphylatoxin inactivator*—cleaves C terminal arginine from C3a and C5a.

Alternate Pathway Activation

–is caused by presence of zymosan, endotoxin, and complexes of aggregated human Ig's, including $F(ab')_2$ fragments.

–bypasses need for specific Ab and early components C1, C4, and C2.

Alternate Pathway Activation Sequence

–low levels of C3b are present in serum under normal circumstances.

–in the presence of Factor D, Factor B can be cleaved into Bb and Ba. Bb will assemble with C3b.

–Properdin P will stabilize C3bBb on the surface of zymosan or endotoxin and allow this entity (P.C3b.Bb) to cleave another molecule of C3 to continue the cascade.

Review Test

COMPLEMENT

DIRECTIONS: For each of the questions or incomplete statements below, *one* or *more* of the answers or completions given is correct. Choose answer:

A = 1, 2, and 3 correct
B = 1 and 3 correct
C = 2 and 4 correct
D = 4 correct
E = all correct

5.1. Complement activation by the "alternate" pathway:

1. will produce anaphylatoxin II but not anaphylatoxin I.
2. can occur in rabbits congenitally deficient in C4.
3. necessitates the activation of C1S to an esterase.
4. can be induced by F(Ab')$_2$ fragments of IgG$_4$, IgA$_1$, IgA$_2$, and IgE.

5.2. Complement is required for:

1. the Arthus reaction.
2. killing by cytotoxic T lymphocytes.
3. the development of glomerulonephritis caused by antigen-antibody complexes.
4. antibody-dependent cell-mediated cytotoxicity.

5.3. Complement:

1. consists of at least 11 different proteins.
2. makes up about 10% of the globulins in normal human sera.
3. is not an immunoglobulin.
4. is not increased in concentration by immunization.

DIRECTIONS: Each group of questions below consists of a series of lettered headings followed by a list of numbered words or phrases. Choose the ONE heading that is most closely related to the numbered words or phrase.

Questions 5.4 through 5.8
A. C3b
B. C4b
C. Both of the above
D. Neither of the above

5.4. Involved in "classical" complement cascade.
5.5. May be Ca^{++} bound.
5.6. Immune adherence.
5.7. Involved in alternate (Properdin) pathway.
5.8. Anaphylatoxin activity.

5.9. The first component of complement (C1) would disassemble into its component parts due to:

A. conditions that have minimized its polymeric behavior.
B. autodigestion.
C. loss of calcium.
D. loss of magnesium.
E. none of the above.

Answers and Explanations

COMPLEMENT

5.1. C. The alternate pathway of C activation starts with cleavage of C3. Thus *both* anaphylatoxin I (C3a) and anaphylatoxin II (C5a) are produced. The alternate pathway does not require the participation of C1, C4, or C2 (i.e., the "early" components) and is induced by F(Ab')$_2$ fragments of almost any of the Igs.

5.2. B. Complement is the key mediator of the glomerulonephritis and vasculitis induced by immune complex (Type III) hypersensitivity reactions. The Arthus reaction is a prototype III reaction.

5.3. E. All of the above statements are true. Complement consists of components C1-C9 plus factors B, D, Properdin, and numerous control proteins. These globulins constitute at least 10% of the total pool, are not immunoglobulins in nature, and have a rather constant concentration in serum.

5.4. C. The classical C cascade is initiated by activation of C1, followed by cleavage of C4 and C2 respectively to yield C3 convertase—C4b2a. C3 is thus cleaved into C3a and C3b.

5.5. D. The C1 components C1q, C1r, and C1s are bound by Ca^{++}. C3 convertase (C4b2a) is held together by Mg^{++}.

5.6. A. C3b receptors are present on macrophages, monocytes, neutrophils, and other immunocompetent cells, and the C3b molecule is thus involved in "immune adherence," which enhances phagocytosis.

5.7. A. The generation of low levels of C3b is the first step in complement activation by the alternate pathway. C4 is an early component and not necessary for the alternate activation scheme.

5.8. D. Anaphylatoxin activity is generated by C3a and C5a. There are specific receptors for these molecules on the surface of most cells.

5.9. C. Calcium holds the C1 components together. Chelation of Ca^{++} will lead to the dissolution of C1 into its C1q, C1r, and C1s subunits.

Hypersensitivity Reactions

Gell and Coombs Classification of Hypersensitivity Reactions

Type I: Anaphylactic Hypersensitivity Reaction—"allergy" reaction; reagin (IgE) mediated.

Type II: Cytotoxic Hypersensitivity Reaction—antigen is cell bound.

Type III: Immune Complex Hypersensitivity Reaction—antigen is soluble.

Type IV: Cell-Mediated or Delayed Type Hypersensitivity Reaction—cell-mediated immunity, *not humoral.*

Type I Hypersensitivity Reaction: Anaphylaxis

–occurs in "atopic" individuals.

–is in response to environmental (i.e., allergens) or administered (e.g., penicillin) antigens.

–is mediated by IgE ("reaginic") antibody bound to surface of *mast cells* or *basophiles.*

–may be localized or systemic.

IgE in Immediate Hypersensitivity

–was first "observed" by Prausnitz and Kustner, who used serum to transfer fish allergy passively to a normal individual.

–was definitely determined by Ishizaka as to structure and function.

–is produced in response to environmental antigens (allergens).

–will bind by Fc portion of ϵ to mast cells or basophiles.

–will cause release of vasoactive and chemotactic factors from mast cell on cross-linking of antigen on surface.

–can be measured *in toto* by use of *r*adio*i*mmuno*s*orbent *t*est (RIST).

–can be measured for specific idiotypes by use of *r*adio*a*llergo*s*orbent *t*est (RAST).

Mast cells, on stimulation of IgE on surface, will release

1. Vasoactive Mediators
   A. Histamine, MW-111—causes smooth muscle contraction in bronchioles and small blood vessels, increased permeability by capillaries.
   B. Platelet Activating Factor (PAF).
   C. Slow-Reacting Substance of Anaphylaxis (SRS-A): metabolites of arachadonic acid, the leukotrienes LTC4, LTD4, and LTE4.
   D. Prostaglandins and Thromboxanes—products of cyclooxygenase metabolism of arachadonic acid.
2. Chemotactic factors
   A. *E*osinophil *C*hemotactic *F*actor of *A*naphylaxis (ECF-A)—MW 1000—cause influx of eosinophils.
   B. *N*eutrophil *C*hemotactic *F*actor—a *h*igh *m*olecular *w*eight (*HMW—NCF*) factor (750,000 MW) for PMNs.

Maneuvers for Treatment of Allergic Reactions

–get rid of allergen or move away from allergen.

–tie up IgE molecules with haptens of allergens or with monovalent antigens.

–*hyposensitization*: inject patient with solubilized allergen to cause production of IgG, which will compete with IgE for allergen.

–*chromalyn sodium* will stabilize the mast cell membrane and decrease the amount of histamine released.

–increase cAMP levels—will stabilize membrane and decrease amount of vasoactive and chemotactic molecules released. This may be done by increase of adenyl cyclase activity (stimulation of β-receptors, isoproterenol), or decrease in phosphodiesterase activity (methylxanthines: aminophylline, theophylline).

Type II Hypersensitivity: Cytotoxic Reactions

Type II reactions involve the production of Ab to specific cell surface epitopes, which will cause destruction of the cell.

Antibody to cell surface antigen can:

–cause reduction in cell surface charges.
–cause opsonic adherence via Fc of Ab to PMNs, macrophages, and K cells; will enhance cell phagocytosis and promote cell death.
–activate complement to cause cell lysis.

Examples of Type II HS: Cytotoxic Rxs

–Transfusion reactions: ABOH incompatibility—IgM versus A or B alloantigens.
–Rh incompatibility—IgG antibodies versus D antigen on fetal red cells.
–Hemolytic anemia—Ab to RBC epitopes.
–Goodpasture's syndrome—Ab to glomerular and bronchial basement membrane.
–Myasthenia gravis—Ab to muscle acetylcholine receptors.

Type III Hypersensitivity: Immune Complex Reactions

Ag-Ab complexes, especially in Ag excess, can cause a series of events that lead to pathologic expression, edema, PMN infiltrate, lesions in blood vessels, and kidney glomeruli.

Ag-Ab complexes can cause:

–platelet aggregation leading to formation of microthrombi and release of vasoactive amines.
–activation of *complement* and release of anaphylatoxins (causing histamine release) and chemotactic factors (for PMNs).
–clotting factor XII activation leading to fibrin, plasmin, and kinin formation.

Examples of immune complex reactions

–Arthus reaction—immunization of rabbits with horse serum (classic prototype of Type III reaction).
–Farmer's lung—Ab to inhaled aspergillus mold.
–Cheesemaker's lung—Ab to fungi.
–Pigeon fancier's disease—Ab to pigeon "dander."
–Serum sickness—Ab to "foreign" immunoglobulin injection.
–Rheumatoid arthritis—Rheumatoid factor (IgM) versus Fc portion of self IgG.

Type IV—Delayed Type Hypersensitivity: T Cell Mediated Immunity

Delayed type hypersensitivity (DTH) reactions are differentiated from immediate-type HS reactions (Types I, II, and III) by the fact that they are examples of cell-mediated

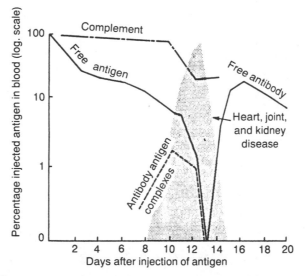

Figure 5.6. Changes in serum constituents during experimental serum sickness in the rabbits induced by a single intraveneous injection of serum protein antigen. The ordinate refers only to percentage antigen concentration and the number of 50 percent hemolytic doses of complement. (Adapted from a diagram kindly supplied by Dr. F. J. Dixon.) (From Brown F, Wilson GS, eds. Topley and Wilson's Principles of Bacteriology, Virology and Immunity. 6th ed. Baltimore: Williams & Wilkins, 1975, vol 2, p 1498.)

immunity (CMI). Types I, II, and III are mediated by Ab and are examples of humoral immunity (HI).

Sequence of Events in a Type IV Reaction: DTH

–An appropriate antigen (tuberculin, PPD, tumor cell, transplanted cell, virally transformed cell) is "processed" by macrophage; epitopes of Ag are expressed on macrophage surface via Class II (Ia) antigens. Mφ produce IL-1.

–T helper cells react to Ag epitope and class II Ag via Ti and CD4, respectively.

–T helper cells are also stimulated by IL-1 from Mφ.

–T helper cells produce IL-2, and IL-2 receptors become fully activated and release lymphokines, having an effect on both B and T cells.

[NOTE: Above sequence is identical in both CMI and HI. CMI will follow the effect of the lymphokines on *macrophages* and *CD8 cells;* HI will follow the effect of these lymphokines on B cells.]

Lymphokines Affecting Macrophages

–Macrophage chemotactic factor.

–Macrophage migration inhibition factor (MIF).

–Macrophage activator factor (MAF); termed interferon γ.

Lymphokines Affecting CD8 Cells

–IL-2 will activate CD^+ cells to become fully CYTOTOXIC.

CD8+ Cytotoxic Cells

–will react to viral and tumor Ag and Class I HLA via Ti and CD8 molecules respectively.

–will be further stimulated by IL-2 from T helper cells.

–will produce IL-2 themselves.

–will produce γ IFN.

Other Lymphokines Produced by CD4 and CD8

TNF – tumor necrosis factor.
OAF – osteoclast activating factor.
HRF – histamine releasing factor.

Examples of DTH and Type IV HS RX

–Tuberculin reaction – Rx to PPD.
–Contact sensitivity – poison ivy
       – poison sumac
       – poison oak
       – patch test
–Allograft rejection.
–GvH reaction.
–Tumor immunity.

Immunologic Disorders

I. Developmental Disorders

–as with any other system, developmental disorders and aberrancies can occur in the diverse cells and molecular signals collectively termed the immune system.

–recognizing and understanding these disorders requires understanding of the normal immune system.

–in normalcy, the immune system is extraordinarily complex, involving multiple, obligatory cellular interactions and molecular signaling; as a consequence of the complexity of the immune system, multiple points exist at which aberrancies might occur.

–contrariwise, multiple sites exist for intelligent interference and exogenous regulation of the system.

–major concerns lie in the immunodeficiency diseases of the neonate and children and the depressed immune response of the aged.

–major diagnostic clue lies in patient's history of multiple infections, which can be verified with quantitative Ig tests, T/B cell ratios, and humoral and cell-mediated responses to common antigens.

A. Transient Hypogammaglobulinemia of Infancy

–reflects a *normal* condition resulting from temporal requirements for full development of the infant's immune system.

–the neonate is born with an adult level of placentally transferred IgG and possesses antibodies associated with the maternal immune experience.

–however, the rate of synthesis of secretable Ig by newborn infants is very low and does not reach adequate levels for a number of months.

–consequently, from generally the third to fifth month, a transient period of physiologic hypogammaglobulinemia exists, beginning with the disappearance of maternally transferred IgG ($t\frac{1}{2}$ = 22 to 28 days) and the onset of significant synthesis of IgG by the infant.

–results in increased susceptibility to some microorganisms.

B. Congenital Agammaglobulinemia (Bruton's)

–sex-linked (male) disorder characterized by recurrent pyogenic infections and digestive tract disorders beginning at 5 to 6 months.

–diagnose by absence of tonsils, germinal centers, B cells, and serum immunoglobulins below 10%.

–defect may lie in the transition from pre-B to B cells since the pre-B cells are normal.

–thymus and cell-mediated immunity appear normal.

–passive transfer of adult serum immunoglobulins protects patient.

C. Common, Variable Hypogammaglobulinemia

–can be acquired at any age by either sex.

–most patients have B cells but do not secrete immunoglobulin.

–depressed serum Ig to less than 250 mg % of IgG and less than 50 mg % IgA and IgM (Normal IgG = 800–1400; IgM = 60–200; IgA = 100–300 mg %).

–present with increased susceptibility to pyogenic infections and autoimmune diseases.

–multiple causes; diverse treatments.

D. Dysgammaglobulinemia

–patients present with a selective Ig deficiency.

–depressed IgA levels (less than 5 mg %) most common (1 in 600–800 individuals).

1. mucosal surface protection is lost
2. patients exhibit normal numbers of IgA cells, but they fail to differentiate into plasma cells
3. increased autoimmunity evident

E. Congenital Thymic Aplasia (Di George Syndrome)

–characterized by absence of T cells, hypocalcemia, and tetany; nonheritable.

–caused by an intrauterine insult to third and fourth pharyngeal pouch resulting in lack of development of thymus and parathyroid between fifth and sixth week of human gestation.

–a depressed cell-mediated immunity permits disease due to opportunistic organisms (e.g., *Candida*, *Pneumocystis*, viral infections).

–vaccination with live vaccines (e.g., measles, small pox) can be fatal.

–germinal centers, plasma cells, and serum Ig appear normal.

–transplantation of fetal thymus is an experimental treatment; however, it may be complicated by a graft versus host reaction.

F. Chronic Mucocutaneous Candidiasis

–a highly specific T-cell disorder characterized by an absence of immunity to *Candida*.

–T-cell absolute numbers and other functions appear normal.

–about half of these patients have endocrine dysfunctions (hypothyroidism).

G. Wiskott-Aldrich Syndrome

–a sex-linked (male) disease with patients presenting with a triad of thrombocytopenia, eczema, and recurrent infections.

–characterized by a depressed cell-mediated immunity and serum IgM, but normal IgG and IgA levels.

–respond poorly to polysaccharide antigens.

–increased lymphoreticular malignancies or lymphomas noted.

–the primary defect may be an absence of specific glycoprotein receptors on both T cells and platelets.

–bone marrow transplantation is an experimental treatment.

H. Severe Combined Immunodeficiency Disease (SCID)

–characterized by a genetic defect in stem cell resulting in absence of thymus, T and B cells.

–half of these patients have a loss in an adenosine deaminase enzyme, resulting in accumulation of toxic deoxyATP, which inhibits ribonucleotide reductase and prevents DNA synthesis.

–results in extreme susceptiblity to infections and a very short life span.

I. Chronic Granulomatous Disease

–characterized by a genetic defect in the nicotinamide adenine dinucleotide phosphate (NADPH) oxidase system.

–results in a defective PMN bactericidal activity due to depressed superoxide dismutase and increased hydrogen peroxide levels.

–laboratory diagnosis based on failure of PMNs and macrophages to reduce a nitroblue tetrazolium dye (NBT).

J. Senescence of the Immune Response

–the aged show a depressed humoral and cellular immune response.

–response highly variable with chronologic age.

–characterized in the main by a loss in some T-cell functions, particularly release of interleukin II and suppressor cells.

–occurrence of autoimmune disease is increased.

Review Test

DEVELOPMENTAL DISORDERS

DIRECTIONS: For each of the questions or incomplete statements below, one or more of the answers or completions given is correct. Choose answer:

      A. if only **1, 2,** and **3** are correct
      B. if only **1** and **3** are correct
      C. if **2** and **4** are correct
      D. if only **4** is correct
      E. if all are correct

5.1. Susceptibility to the yeast *Candida* can occur in:

1. congenital agammaglobulinemia.
2. congenital thymic aplasia (Di George syndrome).
3. common variable hypogammaglobulinemia.
4. chronic mucocutaneous candidiasis.

5.2. The immune deficiency in chronic granulomatous disease is related to the:

1. reduced levels of the fifth component of complement.
2. inability of polymorphonuclear leucocytes to ingest bacteria.
3. dysgammaglobulinemia.
4. inability of polymorphonuclear leucocytes to kill ingested bacteria.

5.3. Severe combined immunodeficiency disease can be characterized by:

1. absence of the thymus.
2. a genetic defect in the nicotinamide adenine dinucleotide phosphate oxidase system.
3. a genetic defect in adenosine deaminase activity.
4. a transient loss in the ability of neonates to express humoral and cell mediated immunity.

5.4. In general, during the aging process:

1. interleukin II levels diminish.
2. suppressor cells increase.
3. the incidence of autoimmunity increases.
4. thymus tissue decreases.

Answers and Explanations

DEVELOPMENTAL DISORDERS

5.1. C. Immunity to *Candida* is predominantly cell mediated, and thus susceptibility is increased also in instances of depressed T-cell function as in the Di George syndrome.

5.2. D. In this disease the ability to kill ingested microorganisms rather than phagocytosis is the dominant lesion. There is no evidence for a loss in complement components or immune globulin.

5.3. B. SCID is a devastating terminal disease involving a lesion in the stem cell population, such that lymphoid tissue does not form. One-half of these patients exhibit a loss in adenosine deaminase activity. A loss in NADPH on the other hand characterizes the PMNs in chronic granulomatous disease.

5.4. E. All statements are correct with the caveat that considerable variability occurs with chronologic age.

Autoimmune Disorders

I. Definition

–disorders in immune regulation resulting in antibody or cell-mediated immunity against the host's own tissues.

–may or may not result in injury to the host.

–normally unresponsive to self-antigens by an unknown mechanism, termed tolerance.

–however B-cell clones do exist in normal human beings with idiotypes reacting with self-antigens.

II. Explanatory Theories

–each may be applicable under different conditions.

A. Microbial antigens cross-reacting with host tissues induce an immune response against self

–not a true autoimmunity as stimulus is of exogenous origin.

–examples: streptococcal antigens cross-react with sarcolemmal heart muscle and kidney.

–anti-DNA antibodies reacting with cells in patients with lupus erythematosus may be induced by microbial DNA.

–deposition of viral antigens on host cell membranes involves an immune reaction against the host cell.

B. Host antigens previously sequestered from fetal tolerance-inducing mechanism are released and become immunogenic

–evidence for such antigens in thyroid and heart tissue that emerge following tissue damage by microbes or surgery.

C. Alteration of host molecules, exposing new antigenic determinants unavailable at time of induction of fetal tolerance

–example: rheumatoid factor in patients with rheumatoid arthritis is mainly an IgM antibody against the Fc fragment of slightly altered IgG.

D. Attachment of foreign hapten to self-molecule formulating a hapten-carrier complex

–example: certain drugs (e.g., quinidine and sulphathiazole) can attach to platelets. Antibody to drug reacts with drug on membrane, activates complement and platelet lyses.

E. Depletion of suppressor cells

–suppression by T suppressor cells of B-cell clones with idiotype specificity for self-antigens under normal conditions may be lost, resulting in autoantibodies.

III. Disorders

A. Examples of Systemic Disorders

1. Systemic Lupus Erythematosus (SLE)

–an episodic multisystem disease usually in young women with vasculitis as a major lesion.

–characterized by multiple autoreactive antibodies, the most dominant of which is antinuclear antibody (ANA).

–cross-reactive ANA may be induced by microbial infection.

–continuous insult by antigen-antibody complexes in antigen excess and complement activation results in nephritis.

–can be confused with rheumatoid arthritis, as 30% of patients exhibit serum rheumatoid factor.

2. Rheumatoid Arthritis

–a chronic, systemic inflammatory disease mainly of the joints.

–characterized by appearance in serum and synovial fluids of rheumatoid factors (antibodies against Ig) and complement activation.

–resulting chemotactic factors attract inflammatory cells into joints, which damage tissues via release of pharmacologically active mediators.

–rheumatoid factor formation may be a response by synovial lymphocytes against microbial antigens.

–a genetic predisposition (HLA-D4 and HLA-DR4) may exist for this condition.

3. Sjögren's Syndrome

–a chronic inflammatory disease primarily of postmenopausal females characterized by autoantibodies against salivary duct antigens.

–patients may present with dryness of mouth, trachea, bronchi, eyes, nose, vagina, skin.

–may occur secondary to rheumatoid arthritis and SLE.

–unknown etiology.

4. Polyarteritis Nodosa

–one of a number of like human vasculitides that can be reproduced experimentally by antigen-antibody complexes.

–complexes of hepatitis B antigen with its specific antibody are found deposited in vessel walls of 30% to 40% of patients.

B. Organ Specific Disorders

1. Blood

–autoantibodies reacting with blood cells to result in anemia; leukopenia and thrombocytopenia can occur (e.g., SLE).

–malignant transformation of a single plasma cell clone (multiple myeloma) results in the appearance of an excess of IgG or other immunoglobulin classes (termed paraproteins).

–such patients may also secrete Bence Jones proteins (monoclonal light chains) in their urine

2. Central Nervous System

(a) Allergic Encephalitis

–a demyelinating disease that can occur following an infection or immunization.

–characterized immunologically by a perivascular mononuclear cell infiltrate in the white matter.

–can be mimicked experimentally by immunization of animals with homologous extracts of brain or a nonapeptide from the basic protein of myelin.

–experimental disease can be transferred with sensitized lymphocytes, thereby implicating cell-mediated immunity as responsible for demyelination.

(b) Multiple Sclerosis

–a chronic, relapsing disease characterized immunologically by mononuclear cell infiltration and demyelinating lesions (plaques) in the white matter of the CNS.

–most patients have increased IgG in cerebrospinal fluid, containing elevated titers to measles and other viruses.

–decrease in suppressor T-cell function indicates an immunoregulatory disorder.

(c) Myasthenia Gravis

–characterized by a defect in neuromuscular transmission resulting in muscle weakness and fatigue.

–associated with presence of an anti-acetylcholine receptor antibody causing loss of receptor.

–thymic hyperplasia or thymoma with increased numbers of B lymphocytes seen in majority of patients.

3. Endocrine

(a) Chronic Thyroiditis

–characterized by autoantibodies and cell-mediated immunity to thyroglobulin or thyroid microsomes

–lesion(s) can be reproduced experimentally by injection of autoantigen with an adjuvant.

–generally a self-limiting disease of females with a likely genetic base.

–tissue damage may occur via antibody-dependent cell-mediated cytotoxicity (ADCC).

(b) Graves' Disease (Hyperthyroidism)

–characterized by autoantibody to the thyroid-stimulating hormone receptor and infiltration of the thyroid gland with T and B cells.

–antibodies may compete with TSH for receptor site and mimic TSH activity.

(c) Diabetes Mellitus

–characterized in Type I diabetes by the destruction of insulin-producing cells through either humoral or cell-mediated anti-islet cell immunity.

–no evidence for autoimmune pathogenesis in Type II.

4. Gastrointestinal Tract

(a) Pernicious Anemia

–characterized by autoantibodies to the gastric parietal cell and intrinsic factor.

–results in inability to absorb vitamin B12.

(b) Ulcerative Colitis

–chronic inflammatory lesion confined to the rectum and colon and characterized by infiltration of monocytes, lymphocytes, and plasma cells.

–patients' lymphocytes show cytotoxicity against colonic epithelial cells in culture.

–antibodies cross-reactive with *Escherichia coli* are present, but disease is of unknown etiology.

(c) Crohn's Disease

–inflammatory, granulomatous disease usually occurring in the submucosal area of the terminal ileum.

–chronic progressive disease, often suspected, but not established, as being of microbial etiology.

(d) Chronic Active Hepatitis

–characterized by infiltration of liver by T and B cells and monocytes.

–suppressor cell numbers are diminished.

–may be a disease of faulty immunoregulation.

Review Test

AUTOIMMUNE DISORDERS

DIRECTIONS: For each of the questions or incomplete statements below, one or more of the answers or completions given is correct. Choose answer:

    A. if only **1, 2,** and **3** are correct
    B. if only **1** and **3** are correct
    C. if **2** and **4** are correct
    D. if only **4** is correct
    E. if all are correct.

5.1. In systemic lupus erythematosus patients:

1. vasculitis is a basic lesion.
2. with nephritis a linear deposition of γ-globulin on their glomerular basement membrane (GBM) occurs.
3. antinuclear antibody is present.
4. rheumatoid factor is rarely present.

5.2. Characteristics of myasthenia gravis include:

1. demyelinating lesions of the white matter of the CNS.
2. thymic hyperplasia.
3. its reproducibility in animals by injection of a nonapeptide isolated from myelin.
4. an anti-acetylcholine receptor antibody.

5.3. The rheumatoid factor *per se:*

1. is the antigen initiating the rheumatoid inflammatory process.
2. is an antibody against cellular DNA.
3. consists of mostly DNA.
4. is an antibody against immunoglobulin.

5.4. In multiple sclerosis patients:

1. increased IgG in the cerebral spinal fluid is typical.
2. mononuclear infiltration occurs in the brain.
3. a decrease in suppressor T-cell function is seen.
4. autoantibodies to intrinsic factor occur with loss of vitamin B12 uptake.

Answers and Explanations

AUTOIMMUNE DISORDERS

5.1. B. Vasculitis and antinuclear antibody are hallmarks of this disorder as is a "lumpy-bumpy" deposition of immune complexes on and behind the GBM. A linear deposition occurs only when the antigen is part of the GBM as in Goodpasture's nephritis. Rheumatoid factors are frequently present in SLE.

5.2. C. Both thymic hyperplasia and loss of actylcholine receptors through antibody-induced capping and involution characterize this disease. It is not to be confused with multiple sclerosis or encephalitis.

5.3. D. Self-explanatory.

5.4. A. Autoantibodies to intrinsic factor characterize pernicious anemia.

INDEX

Page numbers in *italics* denote figures; those followed by "t" denote tables.

231